Business Principles

Landscape Contracting

Business Principles for Landscape Contracting, fully revised and updated in its third edition, is an introduction to the application of business principles of financial management involved in setting up your own landscape contracting business and beginning your professional career. Appealing to students and professionals alike, it will build your knowledge of financial management tools and enable you to relate their applications to real-life business scenarios. Focusing on the importance of proactive financial management, the book serves as a primer for students in landscape architecture, contracting, and management courses and entrepreneurs within the landscape industry preparing to use business principles in practice. Topics covered include:

- Financial management and accountability
- Budget development
- Profitable pricing and estimating
- Project management
- Creating a lean culture
- Personnel management and employee productivity
- Professional development
- Economic sustainability.

Steven Cohan earned his Ph.D. at the Pennsylvania State University, USA. His diverse professional career encompasses plant science research, entrepreneurship as an owner of a retail nursery, and executive management of botanical gardens. He is currently an Emeritus Professor at the University of Maryland, College Park, Maryland, USA, where he served as Program Director for the Landscape Management Program. His course instruction included Financial Applications for the Green Industry, Environmental Horticulture, Internship Seminar, and Green Roofs and Urban Sustainability Seminar.

"*Business Principles for Landscape Contracting* has been a valuable supplement to our curriculum. While our students earn a minor in Business Management, Dr. Cohan's excellent text helps them apply what they learn in their business courses to the Green Industry."

Phil Allen, Certified Landscape Professional,
Brigham Young University, USA

Business Principles for Landscape Contracting

Third Edition

Steven Cohan

London and New York

First edition published by Pearson Education Inc., 2005
Second edition published by Waveland Press Inc., 2013

Third edition published by Routledge
2 Park Square, Milton Park, Abingdon, Oxon OX14 4RN

and by Routledge
711 Third Avenue, New York, NY 10017

Routledge is an imprint of the Taylor & Francis Group, an informa business

© 2018 Steven Cohan

The right of Steven Cohan to be identified as author of this work has been asserted by him in accordance with sections 77 and 78 of the Copyright, Designs and Patents Act 1988.

All rights reserved. No part of this book may be reprinted or reproduced or utilized in any form or by any electronic, mechanical, or other means, now known or hereafter invented, including photocopying and recording, or in any information storage or retrieval system, without permission in writing from the publishers.

Trademark notice: Product or corporate names may be trademarks or registered trademarks, and are used only for identification and explanation without intent to infringe.

British Library Cataloguing-in-Publication Data
A catalogue record for this book is available from the British Library

Library of Congress Cataloging-in-Publication Data
Names: Cohan, Steven M., author.
Title: Business principles for landscape contracting / Steven Cohan.
Other titles: Business principles of landscape contracting
Description: Third edition. | Milton Park, Abingdon, Oxon ; New York,
  NY : Routledge, [2018] | Revision of: Business principles of landscape
  contracting. 2013. 2nd edition. | Includes bibliographical references
  and index.
Identifiers: LCCN 2017042515 | ISBN 9780415788199 (hbk) |
  ISBN 9780415788205 (pbk) | ISBN 9781315225500 (ebk)
Subjects: LCSH: Landscape contracting.
Classification: LCC SB472.55 .C64 2018 | DDC 712.068/4—dc23
LC record available at https://lccn.loc.gov/2017042515

ISBN: 978-0-415-78819-9 (hbk)
ISBN: 978-0-415-78820-5 (pbk)
ISBN: 978-1-315-22550-0 (ebk)

Typeset in News Gothic
by Apex CoVantage, LLC

CONTENTS

Preface		vii
Acknowledgements		ix
1	Structuring for accountability	1
2	Budget development	19
3	Profitable pricing	38
4	Estimating	57
5	Financial management	74
6	Project management	96
7	Financial ratios	108
8	Software applications	128
9	Managing human assets	145
10	Productivity basics	175
11	Creating a lean culture	200
12	Professional development	223
13	Bottom-line leadership	237
14	Economic sustainability	243
Appendices		255
Index		268

PREFACE

The third edition of *Business Principles for Landscape Contracting* finds the landscape industry more stable economically than it was during the last three years of the second edition. The impact on the landscape industry during the economic downturn led to the development of new financial strategies that were essential for companies' sustainability. Several of these strategies are discussed in Chapter 14, "Economic sustainability." Surveys of the landscape companies revealed challenges as well as marketing opportunities. The primary challenge was reducing costs, particularly regarding overhead to offset lower gross margins. Application of the business principles that are discussed in this book enable a company to make those financial adjustments. Further reducing costs while increasing production efficiency has been achieved by a relatively new management concept referred to as lean management. Its application to the landscape industry is discussed in Chapter 11, "Creating a lean culture." In regard to opportunities, commercial landscape construction companies have emerged as general contractors overseeing every operation associated with exterior construction. The challenges of this new role are discussed in Chapter 6, "Project management." Maintenance companies have added new services such as parking lot striping, green roof maintenance, and the incorporation of sustainable landscape management practices. The landscape contracting industry is comprised of companies whose sales range from the hundreds of thousands to over a billion dollars. The challenge these companies face is maximizing production efficiency while maintaining quality production standards, Competitive pricing has added the additional challenge of providing more for less. Regardless of their sales volume, the formula for a company's financial success is the same:

lower direct costs + productivity efficiency = profitability.

This formula can only be implemented if the company identifies, and most importantly measures, each of its production components. This point is articulated in the management philosophy of The former Brickman Group Ltd., "With passion we strive to understand and measure where we are."

This textbook provides a comprehensive description of the financial components that enable a company to measure its bottom-line profitability. The chapters are sequenced in a toolbox fashion, filling the financial management box with essential tools for building a database, a budget, a pricing structure, and an accounting system capable of generating timely reports on the current financial status of the company and its profit centers.

Preface

Since the landscape industry is service oriented, the human element could not be ignored. It's people, not machines, who produce the end product for all landscape companies. Therefore, chapters are devoted to personnel management, professional development, employee productivity, and leadership. The latter is critical to the development and retention of a company's human assets. Integral to the discussion of employee retention is company culture, which is defined and exemplified by both landscape companies and other types of businesses.

This is not a textbook of theory but rather one of practice. Figures and text contain information that have been researched directly from design/build and landscape maintenance companies. Knowledge application exercises at the end of each chapter reinforce the content and its significance to the industry. The knowledge gleaned from the text will benefit students as well as entry-level managers whose career goals include entrepreneurship or management positions within the landscape industry.

ACKNOWLEDGEMENTS

Gratitude is extended to the plethora of industry individuals who provided "real-life" business scenarios for the topics discussed throughout the book. Their advice and critiques were invaluable in the evolution of the text and in establishing its credibility.

I'd like to specifically acknowledge the following individuals whose reviews and contributions had a significant impact on the development of this textbook:

AAA Landscape
  Richard Underwood
Akehurst Landscapes Service
  Ed Delaha
Ariens Company
  Daniel Ariens
  Carol Dilger
  Ron Marciel
Aspire Software
  Kevin Kehoe
  Rob Galloway
Bob Jackson Landscape
  Bob Jackson
  David Welsh
BrightView Landscape
  Bruce Hunt
  Larry Leon
  James Wallace
  Tom Donnelly
Chapel Valley Landscape
  J. Landon Reeve IV
Denison Landscaping
  Josh Denison
Garden Gate Landscape
  Charles Bowers

James River Grounds Maintenance
  Maria Candler
JP Horizons
  James Paluch
Kinnucan Company
  Robert Kinnucan
Live Green Landscapes
  Michael Martin
  Rex Hall
  Brandon Proescher
McHale Landscape
  Kevin McHale
  Phillip Kelly
  Hans Bleinberger
  Ryan Davis
Ruppert Landscape
  Craig Ruppert
  Chris Davitt
  Phil Key
  Jay Long
  Ken Hochkeppel

Acknowledgements

My appreciation is further extended to academic colleagues whose insight and pedagogical perspectives have been incorporated into the manuscript.

Brigham Young University
Dr. Philip Allen
Greg Jolley

Columbus State Community College
Richard Ansley
Steven O'Neal

Hinds Community College
Martha Hill

University of Maryland
Joel Manspeaker

Michigan State University
Dr. Brad Rowe

Structuring for accountability

CHAPTER OBJECTIVES

To gain an understanding of:

1 Accounting methods
 a Distinctions
 b Function
 c Applications
2 Accounting system infrastructure
 a Chart of accounts
 b Guidelines
 c Classification
 d Terminology
 e Accountability

KEY TERMS

accounting system	assets	cash basis method
accounts payable	balance sheet	cash flow
accounts receivable	balance sheet accounts	chart of accounts
accrual method	billings basis	direct costs

Structuring for accountability

direct expenses	inventory	overhead expenses
direct job costs	invoices	percentage of completion basis
expenses	job cost figures	positive cash flow
fixed costs	ledger	profit
G&A overhead costs	liabilities	profit centers
income statement	maximizing profit	retainage monies
income statement accounts	negative cash flow	revenue
indirect costs	net worth or owner's equity	

We are all structured for financial accountability. Our personal budget allocates disbursements of our monthly income to specific categories, such as rent, utilities, mortgage, clothing, entertainment, and so on. Recording these disbursements in the checkbook represents just one facet of accounting, a process that categorizes and summarizes recorded transactions. The checkbook serves as a recording device but does not summarize and categorize the financial entries. We can accomplish the balance of the accounting process either by employing an accounting software program or by manually categorizing transactions in a ledger. A **ledger** is an accounting book that summarizes and categorizes information (credits and debits) from individual accounts into single locations. Information from individual accounts is posted (recorded) on a weekly or monthly basis. Ledgers enable us to easily access this summarized information. Manual entry, however, has been replaced with accounting software programs that enable electronic postings into computerized ledgers. An **accounting system** processes the recorded information into financial statements. These statements provide an assessment of our current financial status and details pertaining to all of the individual financial transactions. We can get breakdowns of our expenses, for example what we spent for entertainment, automobile, clothing, insurance, and so on. How effectively we manage our accounting system and utilize it to make our financial decisions determines whether we will be able to meet our financial obligations.

In the business world, meeting financial obligations and attaining profit goals require astute management of two primary components, cash flow and expenses. **Cash flow** represents the amount of revenue (income) a company receives from the sale of goods and services in relation to its expenses. It can be thought of in terms of the balance between income and expenses. **Positive cash flow** is when the amount of revenue exceeds the amount of expenses, and **negative cash flow** is when expenses exceed revenue within a specific time period. **Expenses** are all the costs associated with the sales of goods and services and those costs associated with business operations, for example for utilities. The successful management of these components cannot occur without a comprehensive accounting system. This system is the source of all the financial information that is derived from recorded transactions. These transactions include all revenue (income) from the sales of goods and services and disbursements associated with the sales and business operations.

Structuring for accountability

Date	P.O.#	Ordered By	For Dept.	Vendor	Job Name	Materials	Inv. #	Yard	Dollars Const.	$ Irrigation Lights	Date Expected

PURCHASE ORDER LEDGER

Figure 1.1 Purchase Order Ledger

Maximizing profit can only be achieved with an accounting system that enables management to monitor the financial status of the company on a daily basis. Unlike monthly financial statements, which are historical documents, daily financial reports provide current information that enables management to identify where adjustments need to be made to meet budget figures. Charlie Bowers, CEO of Garden Gate Landscaping, Inc., summarized the value of such an accounting system: *By being able to see the future and the present earlier, you can be proactive rather than reactive when looking at historical financial statements.*[1]

An example of being proactive would be directing more effort into securing sales in specific profit centers based upon an assessment of actual versus budgeted sales figures. Reviewing current financial data will also enable adjustments to be made in overhead and direct expenses if their percentages are exceeding projections. **Overhead expenses** are those that support the sales and production of goods and services, such as administrative expenses and equipment maintenance. **Direct expenses** are those associated with service production, such as labor and materials.

Once the financial information is categorized and summarized, reports and statements can be generated to provide an analysis of the current financial health of the business. Integrated software programs adapted to the green industry enable immediate access to reports on any facet of the business with the touch of a key. Managers may access current job

Structuring for accountability

cost analyses, a division's current sales and profits against budgeted projections, hardscape inventory, or any other information that has been documented in the system. Some landscape companies provide their managers with tablets or smart phones for accessing the system in the field. Managers utilize this technology to input and download client and job data.

Management depends on current financial information to make decisions on operational procedures, pricing, hiring, equipment purchasing, acquisitions, marketing, and other business activities that require financial transactions. Accounting provides the communication link between management and the business finances. A profile of this communication link is illustrated in Figure 1.2. The output provides management with information that will influence decisions on current production and operations as well as future growth and development.

The professional services of a certified public accountant are essential for addressing tax liabilities and other accounting issues. A CPA analyzes historical financial data and assists in critical tax planning. Some CPAs have backgrounds in financial planning, and can therefore serve as a business consultant. These individuals can also provide feedback on strategic plans. Industry associations, e.g., National Association of Landscape Professionals, and local business groups, e.g., Chamber of Commerce, are resources for locating CPAs. When seeking the services of a CPA, one should ask the following questions:

- Does the CPA have similar clients?
- Who will manage your account?
- What reports will be generated?
- What other services will be provided?

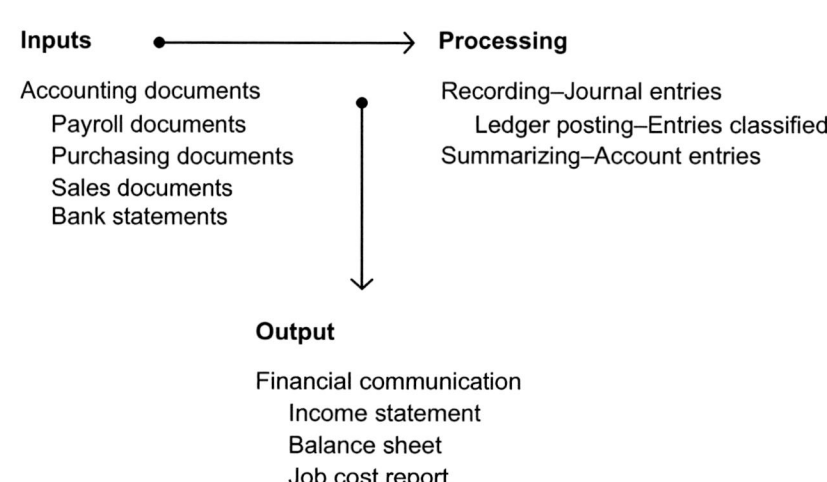

Figure 1.2 Accounting System Profile

ACCOUNTING METHODS

The primary function of an accounting method is to provide management with the most accurate current status of profitability. The different accounting methods are cash basis and accrual basis. Two variations of the **accrual method** are billings and percentage of completion.

The **cash basis method** of accounting records **revenue**, which is the dollar value of materials and services sold, only when payments are received and records expenses only when checks are issued, similar to a checking account. Therefore, it does not reflect the current status of **accounts receivable**, which represents total monies owed to the company by clients or general contractors, or **accounts payable**, which represents total monies owed by the company for job materials. Accounting on this basis inaccurately reflects **profit**, which is the amount of money left after a company has met its current financial obligations.

Inventory, which is purchased materials that are not currently assigned to a specific job (such as boulders, trees, mulch, etc.), is recorded as an expense in cash basis accounting, rather than as an **asset**, which is any item that has monetary value. This assessment is erroneous, since inventory items will ultimately be assigned to contracted projects and will contribute to the total profit of the job. A landscape company can increase its profit by purchasing items at quantity discounts (trees and shrubs, fertilizers, hardscape materials like boulders and fieldstone) and maintaining an inventory to fulfill its seasonal needs. An inventory also can contribute to job cost savings by having items readily available for transport to the job site.

Because of the negatives of cash basis accounting, you may be wondering why a company would use this method. As noted in Table 1.1, it is a simple system and therefore can be implemented easily. It also offers tax advantages since income tax is paid only on the

Table 1.1 Cash Basis Accounting

Advantages:
- Simple to use.
- Offers tax advantage since income can be changed from one financial period to another by regulating collections and expense disbursements.
- Income tax is paid only on the actual money received.

Disadvantages:
- Financial statements misrepresent the current financial status and therefore are useless as a basis for management decisions.
- Most lending institutions will not issue credit based upon financial statements generated from cash-based accounting systems.

Source: *Designing Your Accounting System*, by Ross-Payne & Associates, Inc., National Association of Landscape Professionals (formerly Professional Landcare Network), Herndon, VA

Structuring for accountability

Table 1.2 Distorted Financial Perspective from Cash Basis Accounting

January 1 – March 31, 2017			
Actual		**Cash Basis Accounting**	
Sales revenue	$210,000	Revenue received	$180,000
Direct job costs			
Material	63,000	Materials – paid	51,000
Labor	42,000	Labor expenses – paid	42,000
Total direct costs	105,000	Total direct costs	96,000
Gross profit margin	105,000	Gross profit margin	84,000
Overhead	63,000	Overhead	54,000
Net profit	42,000	Net profit	30,000

revenue received rather than on the total revenue earned. A company's billing schedule can also control the amount of income tax that will be paid within a financial period, according to when the payments are received. Do these tax advantages outweigh the disadvantages of not having an accurate assessment of the company's current financial status? The answer is a resounding no and will be further reinforced in subsequent chapters.

Table 1.2 illustrates how profit can be distorted by the cash basis method of accounting. The $30,000 differential in actual income will be received eventually, but it will be recorded in another financial period. Since overhead expenses are calculated as a percentage of total revenue, this figure is lower with cash basis accounting. In reality, the overhead expenses are based on the percentage of the total income figure. Again, the balance of overhead expenses will be recorded in another financial period.

As previously stated, revenue is only recorded as it is received and expenses as they are paid. Therefore, the cash basis method of accounting never truly reflects a company's income and expenses in a financial period. The same would hold true on individual jobs in which the dollar value of the job and its related expenses would only be reflected as revenue received and expenses paid.

The accrual method of accounting, in contrast to the cash basis, is cumulative, adding income and expenses as clients are billed and invoices are received. **Invoices** are bills issued by vendors for purchases of plants and materials. Two variations of these systems are the billings basis and the percentage of completion basis. The **billings basis** (Table 1.3) records sales with each invoice that is issued and expenses with each vendor invoice that is received. The fact that payments have not been received or that invoices have not been paid are not relevant to this accounting method. Its accuracy, in terms of reflecting profitability, is contingent upon expenses and revenue being accounted for in the same period. If the accounting department is a few weeks behind in posting billings and invoices, generated financial reports will be inaccurate.

Table 1.3 Billings Basis of Accounting

Advantages:
- Simple to use.
- Accurate if job billing is timely.
- Tax liability may be designated for specific periods based upon when job invoices are issued.
- Effective method for short-term installation jobs, which can be completed and invoiced in the same period, and for maintenance contracts.

Disadvantages:
- Long-term installation jobs, which extend over several accounting periods, can result in distorted financial figures.
- If receivables are not collected promptly, the company may pay taxes on uncollected income.

Source: *Designing Your Accounting System*, by Ross-Payne & Associates, Inc., National Association of Landscape Professionals, Herndon, VA

The billings basis of accounting is more effectively employed with landscape maintenance contracts. The revenue associated with these contracts is prorated over twelve months. The monthly invoices (billings) are generated with the same dollar amount regardless of the services rendered during that period. Maintenance clients prefer this form of billing since it facilitates their budgeting process by assessing a fixed monthly amount for landscape services. Landscape maintenance companies are equally satisfied with the system since it provides a fixed amount of cash flow on a monthly basis. Costs associated with the monthly services (labor and materials) are posted as they are incurred (as vendor invoices are received). This method is also applicable for short-term installation/construction jobs that are completed within a financial period.

Accounting systems use financial periods to summarize all transactions. The reason this method is not practical for long-term installation/construction jobs is because they extend over several financial periods. Four of these periods comprise a company's financial or fiscal year.

The accuracy of the billings basis is contingent upon the company's billing schedule, since income is only posted when invoices are issued. Any delayed billing will distort the actual income for a period and thus the company's profit status. The timeliness of billing is equally important on an individual job basis to accurately reflect the job's profit status. In regard to tax liability, a company may pay taxes on uncollected income, since income is recorded based on billings. This situation can be avoided with stringent control over accounts receivable.

The most reliable of the accrual methods for long-term installation/construction contracts is the **percentage of completion basis**. This method is particularly adaptable to the commercial contracting business because it eliminates the inaccuracies associated with timing differences in billings and vendor invoices. The percentage of completion basis accounts for the income and expenses associated with percentage of job completion (e.g., 70%). Billing is submitted to the general contractor, generally on a monthly basis, for payment of services and materials completed to date. This method provides an up-to-date status of the revenue earned and costs incurred throughout the duration of the contract. Upon completion of the entire job, a composite of revenue and costs reported to date will indicate the profit status of the job.

The primary problem associated with the percentage of completion method is that there isn't a standardized technique for assessing the percentage of completion of a job. Therefore, this method is only as accurate as the contractor's method of determining the percentage of completion. However, most professional accountants are in agreement that the percentage of completion method is the most accurate for landscape construction companies.

One of the disadvantages of this accounting method, noted in Table 1.4, is the inclusion of **retainage monies** in the revenue calculation. Retainage monies are those that are retained by the client pending satisfactory completion of the job. Retainage is calculated as a percentage of the total contract price (generally equating to 10% of the total contract price). Since retainage is not paid immediately upon job completion, it is not a component of the revenue received. In essence, it can be considered as part of accounts receivable.

There are advantages and disadvantages to each accounting system: cash, billing, and percentage of completion. Landscape contractors use the billing basis for their maintenance clients and short-term plant installation jobs. Landscape construction and design/build companies that engage in long-term commercial contracts most often use the percentage of completion basis of accounting.

Table 1.4 Percentage of Completion Basis of Accounting

Advantages:
- The most accurate financial profile for installation/construction contracts.
- Allows ample time for tax planning as percentage of completion calculates tax exposure based on completed work.

Disadvantages:
- It requires an accurate method for determining percentage of job completion.
- Retainage monies are calculated into revenue even though they will not be received until a later date.
- This method provides no flexibility for the deferral of income taxes.

Source: *Designing Your Accounting System*, by Ross-Payne & Associates, Inc., National Association of Landscape Professionals, Herndon, VA

Structuring for accountability

CHART OF ACCOUNTS

The **chart of accounts** (Table 1.5) comprises the database from which the accounting system generates financial reports for the business. It provides a list of accounts to categorize all expenses and income associated with the operation of a business. Each account is assigned either an alphabetic or a numeric code to enable classification and access for financial analyses. The information that is categorized essentially becomes a reference library that provides data for the current financial status as well as a financial history. The latter is of particular value for establishing parameters for estimating job cost management since it will have detailed job cost figures. **Job cost figures** encompass labor and material costs and subcontractor costs associated with various facets of design/build, construction, or maintenance jobs. These cost figures are vital to the estimating process of landscape contracting. If the business is new, the owner and accountant may utilize industry job cost standards as guidelines (available through trade association publications).

Since the accuracy of an accounting system depends on the information available, the formatting of accounts should be given careful consideration. Communication between the owner/executive management and the accountant is vital to the immediate and long-term usefulness of the accounting system. As the business expands, the chart of accounts needs to be modified to meet new requirements, for example the addition of an irrigation division or expanding from design/build to maintenance. The accountant must be made aware of the specific information that management requires. It is then the responsibility of management to meet with project managers and accounting staff to determine what information needs to be categorized into specific accounts. Once the accounts are defined, an integrated accounting system can be developed. It should be a system capable of providing a company with comprehensive financial reports. Each contracting business is unique in terms of its market, services, overhead, and so forth. Therefore, the chart of accounts needs to be customized to meet each company's financial information needs.

Table 1.5 Chart of Accounts Guidelines

Develop the chart of accounts according to the nature of the business and the desired financial information.
- Establish a format that can be expanded or adjusted based upon a company's growth and information needs.
- Make sure that accounts clearly define costs and provide the pertinent information required by management.
- Make it concise and user friendly to enhance ease of implementation.

Structuring for accountability

Account coding

A coding system enables a company to categorize financial data and to access category summaries individually or in a financial report format. The classification of accounts is standard in all accounting systems. The coding system may be alphabetic or numeric. Some companies prefer the alphabetic coding system because such a system makes it easier to identify accounts than a numeric coding system. In a numeric coding system (Table 1.6 gives an example), the numbering process begins with a four-digit number that establishes account identification. The first digit represents the account classification, the second two digits are the specific account code number, and the last two digits are a subcode for the respective account. The classification number is specific for each account category; for example, the 1000 series is designated for assets, 2000 for liabilities, and so on. When the numeric codes are being assigned, it is important to allow sufficient room between account designations for future expansion.

Account categories

The chart is comprised of account categories with their respective account listings. Each account category is defined by the individual accounts. The financial data within these accounts provides management with indices for monitoring the account categories. The incorporation of these accounts into financial reports will be discussed and illustrated in Chapter 5, "Financial management."

Balance sheet accounts include all accounts that add financial value to the company and accounts that represent financial obligations of the company. Accounts that comprise a balance sheet include assets and liabilities.

Assets are items that are owned by the company and that have cash value (see Table 1.7):

Current assets – items that can be transformed into cash within a twelve-month period.
Fixed assets – items that have a long-term value, many of which are depreciated (expensed) over a period of years.

Table 1.6 Coding System

4101
4 – account classification, e.g., Income
10 – specific account code, e.g., Installation
1 – subcode for residential

Structuring for accountability

Table 1.7 1000 Asset Accounts

Acct. No.	Acct. Name	Acct. Description
	Current assets	
1001	Cash on premises	Cash funds
1005	Checking accounts	Cumulative cash holdings
1006	Savings accounts	Cumulative cash, e.g., payroll, tax accounts
1011	Investments	Stocks, bonds, certificates of deposit
1030	Accounts receivable – clients	Unpaid invoices, job completions, material sales
1100	Accounts receivable – retainage	Amount of money retained by client, pending job completion
1130	Cost in excess of billing	Value of completed work remaining to be billed
1160	Inventory	Total value of all purchased materials not allocated to jobs
1170	Prepaid expenses	Cumulative payments made prior to due dates, e.g., taxes, insurance, rent
1180	Deposits	Utilities, equipment, rent, etc.
	Fixed assets	
1190	Land	Original purchase price
1200	Buildings	Original purchase price
1210	Furniture and fixtures	Furniture, fixtures, office equipment, based on purchase price
1220	Construction equipment	Purchase price for all equipment
1250	Maintenance equipment	Purchase price for all equipment
1260	Trucks, trailers, automobiles	Purchase price of all vehicles
1300	Improvements	All permanent improvements made to the company's physical facilities
1350	Depreciation – buildings	Depreciation allowance for account 1200
1400	Depreciation – furniture and fixtures	Depreciation allowance for account 1210
1450	Depreciation – construction Equipment	Depreciation allowance for account 1220
1500	Depreciation – maintenance equipment	Depreciation allowance for account 1250
1600	Depreciation – trucks, trailers, autos	Depreciation allowance for account 1260
1700	Depreciation – improvements	Depreciation allowance for account 1350

The descending order in which the asset accounts are listed reflects their degree of liquidity (conversion into cash).

Liabilities are all debts incurred by a company (see Table 1.8):

Current – debts that will be paid within a twelve-month period, e.g., vendor invoices, subcontractors.

Long-term – debts that will be paid beyond a twelve-month period, e.g., loans maturing beyond a year, multi-year equipment leases.

Structuring for accountability

Table 1.8 2000 Current Liabilities

Acct. No.	Acct. Name	Acct. Description
		Current liabilities
2001	Accounts payable – trade	Unpaid invoices from vendors and subcontractors
2030	Accounts payable – retainage	Payments withheld as retainage for subcontractors
2100	Income tax withheld – Federal	Federal payroll taxes
2120	Income tax withheld – State	State payroll taxes
2140	FICA tax withheld	Social security taxes
2160	Notes payable – demand	Loans payable on a date set by the lender, generally short term
2180	Notes payable – current	Loan payments that are made on a monthly basis e.g., vehicles, equipment
2200	Capital leases – current	Lease payments made on equipment, vehicles
2220	Accrued payroll	Earned wages and salaries remaining to be paid
2230	Accrued insurance	Insurance premiums remaining to be paid
2240	Accrued benefit plan	Payments due the company's compensation plan, e.g., profit sharing, pension, pension fund
2260	Accrued taxes – payroll	Total amount of payroll taxes remaining to be paid
2280	Accrued expenses	Additional expenses e.g., rent, sales tax, interest
2310	Warranty reserve	Money withheld from sales for warranty replacements
2230	Accrued taxes – income	Amount due to government agencies for taxes on profits
		Long-term liabilities
2500	Notes payable	Bank loans
2550	Notes payable	Equipment
2660	Notes payable	Mortgages with due dates beyond one year

Table 1.9 3000 Net Worth

Acct. No.	Acct. Name	Acct. Description
3001	Common stock	The original issuance value of all shareholder stock
3010	Retained earnings	The accumulated amount of capital from after-tax profits, since the company's inception
3020	Net profit – current period	After-tax profits from the current year
3030	Dividends	The distribution of after-tax profits to shareholders

Net worth or owner's equity is the cumulative amount of money that is comprised of accumulated company profits, owner's investment, and owner's after-tax profits (see Table 1.9).

The financial data compiled in the balance sheet of accounts enables a balance sheet to be generated. A **balance sheet** reflects the net worth of a company based upon the difference between its total assets and total liabilities. This financial report is an indicator of a company's

Structuring for accountability

Table 1.10 Balance Sheet

Assets	Liabilities & Net Worth
Cash	Accounts payable
Cash in accounts	Notes payable
Securities	Other current liabilities
Accounts receivable – client	Total current liabilities
Accounts receivable – retainage	Long-term liabilities
Inventory	Net worth
Other current assets	Total liabilities & Net worth
Fixed assets	
Total current assets	

Table 1.11 4000 Income

Acct. No.	Acct. Name	Acct. Description
4101	Installation – residential	Revenue from residential design/build installation contracts
4102	Installation – commercial	Commercial build/design installation revenue
4103	Installation – government	Design/build installation revenue from all government contracts
4201	Maintenance – residential	Revenue from residential maintenance contracts
4202	Maintenance – commercial	Revenue from commercial maintenance contracts
4203	Maintenance – government	Revenue from government maintenance contracts
4302	Snow removal – commercial	Revenue from commercial snow removal contracts
4402	Holiday lighting – commercial	Revenue from holiday light installations
4800	Other income	Interest, investments, rental

financial stability, since it shows whether there are sufficient assets to meet the company's financial obligations (liabilities). The format for a balance sheet is illustrated in Table 1.10. The financial significance of the balance sheet will be discussed in further detail in Chapter 5.

Income statement accounts include all the accounts that reflect sources of revenue and **direct costs** associated with generating income from profit centers. **Profit centers** represent production divisions in a landscape company.

A landscape company generally has different production divisions that contract specific services. These divisions are commonly referred to as profit centers. These centers may provide installation, enhancement, construction, irrigation, commercial maintenance, residential maintenance, snow removal, holiday lighting, pest management, pressure washing, and so forth. It is important to chart separate accounts for these profit centers to enable management to analyze their respective profitability. Each profit center will have either a letter or a numerical code to enable generation of profit and loss statements (income statements; see Table 1.11).

Structuring for accountability

Table 1.12 5000 Direct Job Cost Accounts

Acct. No.	Acct. Name	Acct. Description
5100	Plant materials	Subcategories include shrubs, trees, perennials, sod, seasonal color
5200	Hardscape materials	Subcategories include lumber, boulders, gravel, stepping-stones
5300	Freight	Charges associated with plant or hardscape material shipments
5400	Sales tax	All taxes on materials used or sold
5500	Bonds, permits	Costs assigned to respective jobs
5600	Labor	All labor used on the job, inclusive of crew and supervisory personnel
5620	Labor burden	Taxes and benefits for all labor associated with the job
5630	Temporary labor	Labor costs for short-term employees who are hired for specific jobs
5700	Equipment	Costs assigned to jobs based upon time allocated for equipment usage
5800	Equipment rental	Charges assessed to respective jobs
5900	Subcontractors	Expenses associated with all subcontractors contracted on a job

Direct job costs (Table 1.12) are all the expenses directly associated with specific contracts, such as labor and materials, and subcontractors in the company's profit centers. These costs are applied directly to each job in each profit center.

General and administrative (G&A) overhead costs (Table 1.13) are categorized into accounts that are inclusive of all costs associated with supporting the sales and production of goods and services. These costs are generally referred to as **fixed costs** since they remain

Table 1.13 6000 General and Administrative Overhead Costs

Acct. No.	Acct. Name	Acct. Description
6001	Uniforms	Rental costs prorated over duration of job
6010	Replacement materials	Costs associated with materials and labor for warranty items
6020	Inventory variance	Accounts for inventory losses due to plant death, breakage, and theft
6030	Small tools	Costs associated with small tools used on jobs
6040	Equipment, vehicle	All costs associated with vehicle maintenance and equipment maintenance
6050	Fuel and oil	Equipment and vehicle costs
6070	Insurance – Equipment, vehicles	Annual premium costs
6080	Depreciation	Cumulative expense allowance for all owned equipment and vehicles
6090	Leases	Leasing expenses for leased equipment and vehicles
6120	Payroll – mechanics	Salary and benefits
6130	Licenses and fees	Licenses and permit fees for vehicles, trailers
6150	Shop tools, equipment, supplies	Costs associated with shop operation and maintenance
6180	Shop parts inventory	Parts used for equipment and vehicle maintenance
6200	Officer salaries and benefits	Company/corporate officers
6210	Manager salaries	All salaried managers
6220	Clerical salaries and benefits	Administrative staff

Structuring for accountability

Acct. No.	Acct. Name	Acct. Description
6250	Bonuses	Salary incentives/merit increases
6260	Professional services	Legal counsel, payroll processing, accounting, consultation
6280	Utilities	Office, cell phones, radio systems
6370	Maintenance – exterior	All buildings and grounds
6400	Maintenance – office	Office equipment including maintenance contracts
6420	Professional development	Training seminars, conferences, class registration fees, training tapes
6450	Memberships, subscriptions	Professional association dues and publications
6470	Property insurance	All facilities, equipment, liability, theft
6490	Vehicle insurance	All company road-operated vehicles
6500	Marketing	Advertising, promotional items
6520	Donations	Financial and material contributions

Table 1.14 Maintenance Cost Accounts

Acct. No.	Acct. Description
5102	Plant materials
5602	Labor
5622	Labor burden
5702	Equipment
5902	Subcontractors
5903	Fertilizers/chemicals

relatively the same during changes in sales volume. Included in this category of accounts are those referred to as **indirect costs**, costs associated with jobs but not involved in the production aspect of jobs, such as uniforms or small tools.

An illustration of how direct costs may be accessed for a maintenance division analysis is shown in Table 1.14. The code indicates directs costs designated by 5, with the second digit for the respective cost accounts within this category. The profit center of maintenance is designated by the 2 in the code. These accounts would be broken down further for analysis with subcodes designating residential, commercial, and government contracts. Companies with multiple branches also assign an alphabetical or numerical subcode for the respective branch, for example B or 3 for Baltimore, A or 4 for Alexandria, and so on.

5102–0203
Direct cost, plant materials, maintenance, commercial, Baltimore branch

An income statement format that incorporates the financial data contained with the income statement accounts is illustrated in Table 1.15. The **income statement** is also

Table 1.15 Income Statement

Earned revenue
Net earned revenue
Direct job costs
Direct labor
Direct labor burden
Plant materials
Hardscape materials
Material taxes
Subcontractors
Equipment
Other direct job costs
Total direct job costs
Gross margin
Indirect overhead
Indirect labor
Replacement expenses
Small tools and supplies
Equipment rental
Fuel and oil
Other indirect costs
Total indirect overhead
General & administrative overhead
Advertising
Depreciation
Insurance – health
Insurance – vehicles
Insurance – liability
Insurance – workers' compensation
Office expense
Utilities
Salaries – officers/owners
Salaries – clerical and administrative
Salaries – sales & commissions
Travel & entertainment
Recruitment
Payroll taxes
Other G&A overhead
Total general and administrative overhead
Operating profit
Other income – interest
Other expenses
Net profit – before taxes
Net profit + owner's salary
Profit

referred to as a profit and loss statement, since it summarizes current revenue, direct and indirect costs, and overhead. This financial report can be generated on a corporate level, for the production division (profit center), or for an individual job.

The chart of accounts presented in this chapter represents the accounting infrastructure that is necessary to build a database for financial management. The number of accounts will vary depending on the amount of financial information that management decides it needs to access. One company may account for plant materials under one account, while another may have subcategories for trees, shrubs, groundcovers, and so on. Its maintenance division may have one account for materials while another company may want to access information on specific materials, such as fertilizers and pesticides. Subsequent chapters will illustrate how the data in the chart of accounts is applied to budget development and financial analyses.

SUMMARY

Accounting is a process that records all financial transactions associated with a business. The recorded financial data is classified into income or expenses and posted into account ledgers. The posted data enables management to access the current financial status of individual accounts. The recording and posting of financial data provides the database for generating financial reports. Management's analysis of these reports enables it to determine the current financial status of a company and to render decisions, which will enhance its profitability.

What are the primary distinctions between accounting methods?

The cash basis of accounting is similar to a checking account with respect to the recording of transactions. This method only records income upon receipt of payments and expenses only as invoices are paid. Therefore, the cash basis of accounting does not reflect the actual profitability of a job since the incomes and expenses are distorted. This system does not provide management with a resource for making financial decisions.

The accrual basis of accounting is a cumulative system, which records income and expenses as client invoices are issued and vendor invoices are received. The two variations of the accrual method are the billings basis and percentage of completion. The billings basis provides an accurate depiction of income and expenses for short-term jobs in which the client invoices and vendor invoices are recorded within the same period. A more accurate method for reflecting the financial status of an installation job is with the percentage of completion basis of accounting. As each phase of the job is invoiced, all expenses are accounted for on a percentage basis, resulting in 100% accounting for all income and expenses upon job

Structuring for accountability

completion. The billings basis accounting method works well with landscape maintenance contracts, because total annual costs are prorated over a twelve-month period and billed accordingly. Since the costs are predetermined, an accurate depiction of the contracts financial status is achieved by comparing actual monthly posted costs to the budgeted costs.

What is the function of a chart of accounts?

A chart of accounts is analogous to a business library system. Within the system is financial information, which pertains to every aspect of the business's operations. Management establishes accounts in collaboration with the accountant to provide a database from which financial reports may be derived. In addition, the chart of accounts is vital as a financial history for estimating jobs and monitoring production efficiency. This database of financial information provides a comprehensive profile of the company's net worth and profitability with its integration into balance sheets and income statements.

KNOWLEDGE APPLICATION

1. Choose a retail garden center or grocery store in your area. After surveying the facilities and inventory, establish a chart of accounts for revenue profit centers (departments) and for associated direct costs.
2. Using your checkbook, set up a personal ledger sheet with columns for each expense category: for example, books, entertainment, gas, insurance, and so forth. Enter totals for an academic year. Total the columns and calculate the percentage of your total expenses each category represents.
3. Review the liability accounts in the balance sheet of accounts. Select four of these accounts and explain how they might have a negative impact on the financial stability of a company.

NOTE

1. Charles Bowers, 2004, CEO Garden Gate Landscaping Inc., Personal Communication.

2

Budget development

CHAPTER OBJECTIVES

To gain an understanding of:

1. The budget development process
   a. Forecasting
   b. Zero-based budgeting
2. Budgets as financial tools
   a. Operating budgets
   b. Capital budgets
   c. Cash flow budgets
3. Profit and why it is the starting point for budget development
   a. Retirement of debt
   b. Equipment purchases
   c. Working capital – growth
   d. Employee profit sharing, bonuses
4. Budget management as an essential element in financial stability
   a. Financial blueprint
   b. Daily financial report card
   c. Resource for proactive management
5. Management reports and interpreting budget variances
   a. Income statements (profit and loss)
      1. Profit centers

Budget development

KEY TERMS

account manager	depreciation	profit centers
add-ons	forecasting	strategic plans
asset capital	gross margin	watchdogs
budget	net profit	working capital
capital budgets	operating budget	working capital requirement
cash flow budgets	principal	zero-based budgeting
cost structure	profit	

A **budget** is a financial game plan for the future that projects what revenue will be generated and what expenses will be incurred over a specific period of time. On an individual basis, it would include salary income and expenses, including housing, entertainment, food, automobile expenses, and so on.

The projected schedule of the ongoing expenses of running a business in combination with expected income revenues is referred to as an **operating budget**. It represents an educated guess based upon historical financial information. These projections can then be incorporated into a spreadsheet, which at a glance will reflect positive (revenue exceeds expenses) or negative (expenses exceed revenue) cash flow on a monthly basis. The budget projections enable a company to anticipate when adjustments need to be made to ensure that there will be sufficient funds to meet projected expenses of operation. These adjustments may require transferring funds from savings, obtaining a loan, or deferring certain purchases. It is therefore evident that a budget is an important financial management tool for running a business successfully, as well as a financial barometer.

In the business environment, budget development is vital to a company's financial health, growth, and development. If it is managed astutely, the company will be profitable, enabling it to reinvest its profits for future growth, retire debt, provide additional compensation for employees, and fulfill its tax liability. **Profit** is the amount of money that is left after a company has met all of its current financial obligations:

- Debt payments – bank loans, etc.
- Equipment purchases
- Working capital – cash for projected growth
- Bonuses, profit sharing, pension contributions
- Tax liabilities – city, county, state, federal

An operating budget relies on financial data from previous years for its projections of revenues and expenses. Since a budget is considered at best an educated guess, it builds a

Budget development

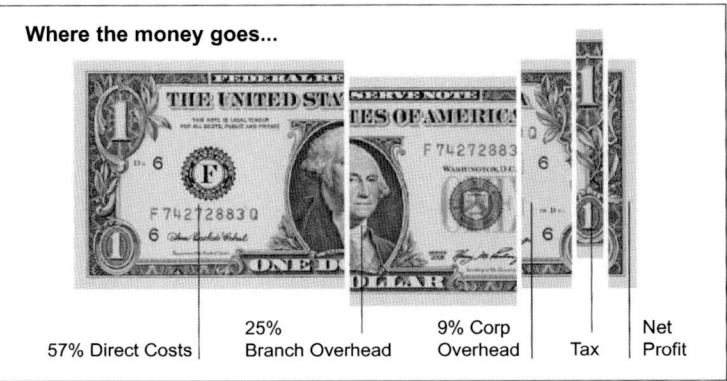

Figure 2.1 Profit Allocation

structure in the form of a financial blueprint for management to revisit on a regular basis. The budget is the most important tool in the financial toolbox. Managers can compare actual financial figures against the budgeted amounts and determine where adjustments need to be made to stay on target for the company's projected profit. Just as a company's budget is its daily financial report card, budget projections enable upper management to communicate the current financial status of the company's operations to branch or division managers. Adjustments then can be made for variables such as weather, the economy, and new sales opportunities.

Other budgets utilized by companies include:

- **Cash flow budgets**, which project cash balances over a specific period: monthly, quarterly, or annually. These budgets alert management to the need to secure loans or to apply cash reserves during periods when there are cash shortages (due to expenses exceeding revenue).
- **Capital budgets**, which are developed to allocate money for purchases of assets, such as vehicles and equipment.

The discussion in this chapter is directed toward operating budgets, since they are a company's primary financial management tool.

The operating budget development process is illustrated in Figure 2.2. As illustrated, the budget is profit driven. Key components of this process are costs – direct, indirect, and overhead – derived from the chart of accounts. Calculated as a percentage of sales, these costs provide parameters for establishing budgets. When sales are projected for subsequent years, the cost parameters are applied to establish corresponding costs. Additional components include revenue/sales projections and the requirements for the working capital needed to support increased sales, which relates primarily to expenses for additional personnel and

Budget development

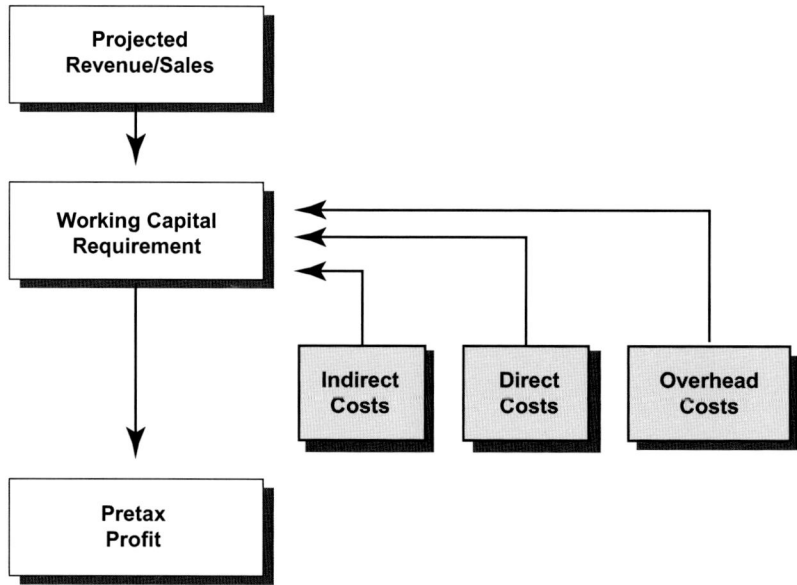

Figure 2.2 Operating Budget Development Flowchart

equipment. The forecasting discussion that follows explains how capital requirements are determined for projected increases in revenues.

Each of the budget development components is critical to the end product, which serves as a comprehensive financial plan for the future. The budget serves as a barometer, enabling management to track whether the company is on target to meet its projected profit.

Forecasting is a process that determines the working capital, asset, and personnel needs for projected revenues. These determinations are based upon financial data in the company's database. In the example that follows, a forecast is developed for the BradCo Landscape Company, a commercial maintenance company. Its financial data (Table 2.1) includes the following:

The cost percentages were derived from last year's income statement (Table 2.2). They represent their respective percentage of the total company's sales; for example 54% direct costs ($1,485,000/$2,750,000). Before these cost parameters are utilized in the forecasting process, the percentages need to be validated as being representative of previous years. New businesses can obtain guidelines for the above benchmarks by referring to the *Operating Cost Study* (2012) compiled by the National Association of Landscape Professionals. This study profiles landscape contracting businesses within specific sales volumes and their respective costs of operation. The figures that will be used in the subsequent company example are within the range of those noted in the industry survey.

Table 2.1 BradCo Financial Data

Costs based on percent of sales:	
Direct	54%
Indirect	10%
Overhead	30%
Financial ratios:	
Sales/fixed assets	10:1
Sales/working capital	9:1
Marketing ratios:	
Maintenance contract retention	90%
Additional maintenance sales	$.50 per dollar of contracted sales
New sales per salesperson	$300,000
Company ratios:	
Sales per full-time employee (FTE)	$65,000
Employee (field) turnover	30%
Business per account manager	$750,000
Business per maintenance crew	$250,000
Administrative staff to $1 million in sales	1.5:1

Table 2.2 BradCo Residential Maintenance 2017 Income Statement

RESIDENTIAL MAINTENANCE 2018 INCOME STATEMENT

Net sales	**$2,750,000**	**100%**
Direct job costs		
Direct labor	825,000	30.0
Direct labor burden	82,500	3.0
Material costs	440,000	16.0
Subcontractors	110,000	4.0
Other	27,500	1.0
Total direct job costs	**$1,485,000**	**54%**
Gross profit margin	**$1,265,000**	**46%**
Indirect costs		
Indirect labor	96,250	3.5
Replacement expenses	8,250	.3
Small tools & supplies	55,000	2.0
Equipment rental	60,500	2.2
Fuel & oil	55,000	2.0
Total indirect job costs	**$275,000**	**10%**

(Continued)

Budget development

Table 2.2   (Continued)

General & administrative overhead		
Advertising	27,500	1.0
Depreciation	82,500	3.0
Insurance – hospital	41,250	1.5
Insurance – liability	27,500	1.0
Insurance – workers' compensation	41,250	1.5
Office expense	27,500	1.0
Payroll taxes	27,500	1.0
Profit sharing	13,750	.5
Lease/land, facilities	82,500	3.0
Salaries/owners	192,500	7.0
Salaries/administrative	137,500	5.0
Salaries/sales	82,500	.3
Telephone/radios/office equipment	33,000	1.2
Utilities	8,250	.3
Other	74,250	2.7
Total G&A overhead	**$825,000**	**30%**
Net pretax profit	**$165,000**	**6%**

Table 2.3   BradCo Net Profit Forecast

	Current	Next Year	Increase
Sales	$2,750,000	$3,300,000	20%
Net pretax profit	**165,000**	**330,000**	**4%**

The management of the BradCo Landscape Company has determined that it can increase next year's sales by 20% and improve bottom line net profit from 6% to 10%. **Net profit** is the amount of money that remains from revenues when all costs (direct, indirect, and overhead) have been paid and before taxes are paid.

To meet the revenue and profit goals, the corresponding working capital, personnel costs, and sales and asset requirements must be determined. The five-step forecasting process, proposed by Kevin Kehoe, business consultant, illustrates how these determinations are calculated:

I. Sales and net profit determination

A projected sales increase of 20% over last year's sales is $3,300,000 ($2,750,000 + $550,000). The projected net profit of 10% for next year's sales of $3,300,000 is $330,000.

Budget development

II. Determination of working capital requirement

Using the current year's sales and working capital figures, $2,750,000/305,555, a sales/working capital ratio of 9:1 is established. **Working capital** is the amount of money that is required to support projected growth (increased revenues).

Table 2.4 BradCo Working Capital and Fixed Assets Forecast

	Current	Next Year	Increase
Sales	$2,750,000	$3,300,000	$550,000
Working capital	305,555	366,666	61,111
Fixed assets	275,000	330,000	55,000

This ratio indicates that BradCo Landscape Company can generate $9 in sales for every $1 invested in working capital. Applying this ratio to next year's sales increase of $550,000 indicates that $366,666 ($3,300,000/9) of working capital will be required to support next year's sales. Therefore, an additional $61,111 of working capital will be required to support the projected revenue of $3,300,000. The sales/fixed assets ratio of 10:1 ($2,750,000/$275,000) represents an additional $1 of assets required for every $10 increase in sales. Based upon next year's projected sales increase, an additional amount of $55,000 ($550,000/10) is required for fixed assets.

III. Determination of new sales requirement

Once the sales volume has been projected, the next step in the forecasting process is determining what the sales requirement will be to meet the 20% increase. The starting point for this determination is the maintenance company's contract retention rate. Last year, BradCo's total maintenance division sales were $2,750,000. This sales figure represents contract sales of $2,000,000 and $750,000 of add-ons ($.375 per contract dollar). **Add-ons** represent additional sales to the client beyond the current contract, such as landscape renovations or seasonal color. Based upon a 90% contract retention (based on contract renewals), the BradCo Landscape Company's maintenance contract sales for next year are projected at $1,800,000 (90% of $2,000,000) plus add-on sales of $675,000 ($1,800,000 × .375), for a total of $2,475,000. In order to meet the company's projected sales goal of $3,300,000, an additional $825,000 ($3,300,000 − $2,475,000) of new contracts will need to be sold.

Budget development

Table 2.5 BradCo New Sales Requirement Forecast

	Current	Next Year
Contract sales	$2,000,000	$1,800,000
Add-ons	750,000	675,000
New contract sales		825,000
Total sales	$2,750,000	$3,300,000

Table 2.6 BradCo Personnel Forecast

	Current	Next Year
Sales	$2,750,000	$3,300,000
Sales per employee	59,782	59,782
Number FTEs	46	55
Turnover rate	30%	
Recruiting needs	20	23
Administrative needs	1	1 full time, 1 part time
Additional account managers	3	4

IV. Determination of personnel requirements

One of the greatest challenges facing the landscape industry is maintaining a sufficient labor force to meet its current and future needs. When sales projections are made, this challenge is further accentuated. The fourth step in the forecasting process determines personnel requirements for the projected revenues.

Determination of production personnel requirements starts with totaling the production hours for all profit centers. **Profit centers** are revenue-generating divisions, e.g., maintenance, installation, irrigation, etc. Only the hours of FTEs (full-time employees) are included in the total. FTEs are production personnel who have worked 2,100 hours per year (the total varies by region). Of the 2,100 hours, 1,800 have been documented, in our example, to be production related. The total current year's production hours for the BradCo Landscape Company are 82,800 (1,800 × 46 FTEs). The sales per FTE were calculated to be $59,978 ($2,750,000/82,800 hours = $32.21/hour × 1,800 production hours). Applying this sales/FTE to next year's forecast determines that 55 FTEs ($3,300,000/$59,978) are required to generate $3,300,000 in total production sales.

Next year's sales require 9 more FTEs than the previous year. In addition to recruiting 9 production personnel, the BradCo Landscape Company needs to hire an additional 14 employees to account for employee turnover (.30 × 46). Therefore, the total number of FTE production personnel to be recruited for next year is 23 (14 + 9).

Budget development

Administrative support is critical to the support of increased sales, particularly from the perspective of contract management and accounting. Industry surveys indicate that one administrative staff employee is required for every 1.5 million dollars in construction revenues and 2 million dollars in maintenance. Based upon BradCo's next year's projected revenues, one full-time and one part-time administrative employee are required.

The example below represents the actual administrative personnel allocation for a Mid-Atlantic landscape company's construction and maintenance divisions:

Table 2.7 Mid-Atlantic Landscape Construction Division (anonymous)

Construction Division
- 1 Branch administrator
- 1 Contract administrator
- 3 Estimators

5 Total administrative personnel
4 Branches = 20 administrative personnel + 7 corporate allocated* = 27
Revenue: $40,500,000 ($10.125 million/branch)
$1,500,000 revenue per administrative employee

Maintenance Division
- 1 Branch administrator

1 Total administrative person
11 Branches = 11 administrative personnel + 10 corporate allocated* = 21
Revenue: $42,000,000 ($3.818 million/branch)
$2,000,000 revenue per administrative person

* Training & recruiting, safety, public relations, fleet, accounting, payroll & benefits, reception

Another key personnel area that needs to be addressed is account managers. **Account managers** (of maintenance companies) are individuals who manage company contracts. The amount of revenue that an account manager is capable of managing effectively, particularly regarding maintaining client relations and production standards, depends on several variables. These include the size and services of the maintenance sites, supervisor proficiency levels, and distances between client properties. The company in our example has assessed the amount of revenue per account manager to be $800,000. On this basis, its next year's increased revenue projection of $550,000 does not warrant hiring an additional account manager, but it may justify grooming an individual in an assistant account manager position for future promotion.

V. Budget parameters

The final step in the planning process establishes the budget parameters for all the expenses associated with the company's projected revenues. The company's financial ratios (costs/sales) are used to establish these financial parameters.

Budget development

Table 2.8 Budget Parameters

Sales	$3,300,000	100%
Direct costs	1,782,000	54%
Indirect Costs	330,000	10%
Overhead G&A total costs (G&A) operating	**858,000**	**26%**
Net Pretax Profit	**330,000**	**10%**

The financial information detailed above now enables the BradCo Landscape Company to develop an operating budget that incorporates the expenses associated with the projected growth. The company's working capital budget will incorporate the amount of money required for purchase of assets (e.g., equipment) to support next year's revenue projections. Management's budget discussions will therefore be based upon realistic expectations rather than suppositions based upon anticipated needs.

THE BUDGET DEVELOPMENT PROCESS

The preceding forecasting process initiated the framing of the operating budget. The remaining infrastructure is built upon the best educated guess scenario within divisions and branches regarding their respective budgets. The finalized company budget expresses in dollars, on a month-to-month basis, the progression of revenues and expenses for the entire company. As previously stated, budgets are profit driven, and therefore, cost accountability is critical to ensure attaining the company's profit goal. The financial data in the chart of accounts collectively become the building blocks for constructing a budget.

Simplification of the budgeting process can be detrimental to a company because of the inaccuracies that then become manifest in the final budget. Budgeting methods that are indicative of such inaccuracies include the following:

- **Growth rate method** – projects revenue increase based on last year's growth, making the erroneous assumption that variables will remain constant. Personnel and equipment needs are not taken into account to meet the revenue goal.
- **Percentage method** – a method that selects a desired percentage of revenue increase and uses it as a parameter for estimating costs. The fallacy of this method is that it assumes costs will increase proportionally to the percentage of increased revenues.
- **Inflation rate** – this method reflects back on last year's inflation rate and adjusts budgeted expenses by the respective rate. As with the percentage method, costs cannot be equated to an external factor.

Table 2.9 Profit Requirement Determination

• Principal payments on debt	$62,600
• Fixed asset acquisitions	113,310
• Cash/working capital required	61,111
• Estimated excess compensation	58,000
• Income tax liability	66,000
Total capital requirement	**$361,021**
• Estimated depreciation	(113,964)
• Profit Requirement	$247,057

A misconception among those who use the growth rate and percentage basis methodology is that revenues are the primary entity that drives the budgeting process. Have you ever heard a person say that he or she is in business to generate revenues? Or that an investor selects a company's stock solely on its revenues? The answer is an emphatic no. Profit is what businesses strive for and depend upon for their growth. A key point to remember is that *increased revenues do not necessarily generate increased profit*.

Profit, therefore, becomes the starting point for the budgeting process. The BradCo Landscape Company management determines how much profit is needed based upon the items in Table 2.9.

The profit requirement determination begins with the principal associated with debt that must be paid within the year. The **principal** is the total dollar amount of a loan without interest. The interest is not included in this calculation, since it is budgeted over the twelve months as an overhead expense. Principal associated with debt is separated into that which is associated with bank loans, $62,600 in this case, and that of fixed asset acquisitions, $113,310.

When reference is made to fixed assets, the asset in question is primarily equipment. The revenue projection stipulated in the BradCo Landscape Company example was $3,300,000, which represented a $550,000 increase in revenues. Budgeting the **working capital requirement** to support the operational expenses, associated with the projected growth in revenue, entails dividing the dollar increase by a factor of 9, as obtained from the *2012 Operational Cost Study of the Green Industry* (National Association of Landscape Professionals, Herndon, VA). Therefore, the sales growth projection of $550,000 has a capital requirement of $61,111.

An additional profit need determination of $58,000 must be made for excess compensation, bonuses, profit sharing, or pension plans, since there is no other source of funding for this expense.

Allocation of profit for tax liabilities is another determination entity, one that will vary from state to state. In the example, a 20% estimate for state, county, and federal taxes on projected profit amounts to a tax liability of $66,000.

Asset capital is the amount of money that is needed to purchase assets for supporting the projected revenue increase.

Budget development

The total capital requirement for the BradCo Landscape Company to meet its financial liabilities, support growth, and provide for excess compensation is $361,021. Deducted from this amount is depreciation of $113,964. **Depreciation** is a tax allowance for the decreased value of equipment and vehicles. The balance of $247,057 is the minimum amount of profit needed to meet the company's financial liabilities. Based upon BradCo's projected sales of $3,300,000, a net profit of 7.5% is required to meet its financial commitments.

Budgeting general and administrative overhead expenses is the next step in the process. The accounts associated with this category of expenses are listed in the company's 2016 Income Statement (see Table 2.2). They comprise all of the expenses associated with the cost of doing business: administrative salaries, professional development, equipment maintenance, and so forth. Management personnel assess each of these accounts on the dollar amount needed to support projected revenues. Therefore, last year's figures or percentages are not automatically inserted into the next year's budget. Instead, it is advisable for companies to use zero-based budgeting. **Zero-based budgeting** projects account costs based on future needs rather than on historical data. On this basis, each expense must be justified in regard to how it helps to achieve the company's financial goals. In the budgeting discussions, managers may, for example, decide that the $50,000 equipment parts inventory can be reduced, since parts are readily available from local sources. Since overhead costs account for approximately 30%, management tries to reduce expenses wherever feasible.

Further consideration is also given to strategic plan goals that are to be implemented in the coming year. **Strategic plans** specify goals and strategies to achieve them within a specific period of time, generally three to five years. One of these goals may be to increase residential design/build sales through participation in a local garden show. The cost of this involvement would be added to the advertising budget.

The overhead expense sheet is a useful tool for analyzing accounts as to how well they support next year's projected revenues. Prior to budgeting the account expense, management does an analysis of the company's advertising program and its current return (sales generated) on invested dollars. Since referrals are a primary source of new business for landscape companies, all advertising dollars need to be justified on the basis of tracking new business with the respective advertising venues. Management decided, on the basis of their analysis, that the *Maryland Living* magazine ads provided name recognition but did not warrant more than four ads. In addition, after two years of tracking referrals from participation in showcase home exhibits, it was decided that the associated expense was not warranted. After considering feedback from industry colleagues regarding job referrals and attendance at a showcase home show, a decision was made to reallocate those dollars to the local garden show.

Management often will assign **watchdogs** to monitor specific overhead accounts throughout the budget year (Table 2.11). The assigned managers ensure that the budgeted expense for their respective accounts does not deviate from its projection.

Table 2.10 Overhead Advertising Expense Sheet

 **ADVERTISING EXPENSES**

Division: Enhancements
Budget Year: 2018
Account Number: 7000
Account Title: Advertising
Descriptions: All expenses associated with enhancing the sales of residential installation such as brochures, ads, and garden show displays are included in this account.

Cost Projections: Expenses	Quantity	Unit Cost	Total Cost
Ads – *Maryland Living*	4	$1,200	$4,800
Garden show displays	1	30,000	30,000
Miscellaneous ads	4	250	1,000
Total: Last year $28,200			$35,800

Monthly Advertising Allocation:

Jan.	$1,200	Feb.	$30,000	March	$1,200	April	$1,450		
May	$250	June	$250	July	0	August	0		
Sept.	$250	Oct.	$1,200	Nov.	0	Dec.	0		

Table 2.11 Watchdog Account Management

 **ACCOUNT MANAGEMENT**

Division: Maintenance
Budget Year: 2018
Watchdog: Avery
Account Number: 6030
Account Title: Small tools
Annual Projections:

Item	Quantity	Unit Cost	Total Cost
Grass rakes	30	$15	$450
Long-handled shovels	20	18	360
Trowels	50	5	250
Steel-tine rakes	20	18	360
Gloves	60	8	480
2014 Total $1,765			$1,900

Budget development

A twelve-month spreadsheet of the budgeted overhead expenses enables close monitoring with the actual expenses incurred. On a month-to-month basis, any indication of deviations from the budgeted expenses will enable management to determine where adjustments need to be made to stay on target to meet the budgeted figures.

The next step in finalizing an operating budget is to combine the total annual expenses (Table 2.12). In essence, management generates a projected income statement, itemizing

Table 2.12 BradCo Landscaping Company 2018 Projected Income Statement

2018 PROJECTED INCOME STATEMENT

Net sales	**$3,300,00**	**100%**
Direct job costs		
Direct labor	990,000	30.0
Direct labor burden	99,000	3.0
Material costs	528,000	16.0
Subcontractors	132,000	4.0
Other	33,000	1.0
Total direct job costs	**$1,782,000**	**54%**
Gross profit margin	**$1,518,000**	**46%**
Indirect costs		
Indirect labor	49,500	1.5
Warranty expenses	9,900	.3
Small tools & supplies	33,000	1.0
Equipment rental	39,600	1.2
Fuel & oil	66,000	2.0
Other	66,000	2.0
Total indirect job costs	**$264,000**	**8%**
General & administrative overhead		
Advertising	33,000	1.0
Depreciation	99,000	3.0
Insurance – hospital	49,500	1.5
Insurance – liability	33,000	1.0
Insurance – workers' compensation	49,500	1.5
Office expense	33,000	1.0
Payroll taxes	33,000	1.0
Profit sharing	16,500	.5
Lease/land, facilities	99,000	3.0
Salaries/owners	231,000	7.0
Salaries/administrative	165,000	5.0
Salaries/sales	9,900	.3
Telephone/radios/office equipment	39,600	1.2
Utilities	9,900	.3
Other	23,100	2.7
Total G&A overhead	**$924,000**	**28%**
Net pretax profit	**$330,000**	**10%**

Budget development

expenses as a percentage of projected revenue. In some cases, financial history is utilized if the expenses as a percentage of sales have been consistent. The balance of expenses is projected based on management's assessment of need (zero-based budgeting).

This financial data is then distributed over a twelve-month period based on trends established from historical data (monthly revenue and expenses). The finalized monthly income statement becomes a working document, which is revisited on a daily basis. If there appears to be a major deviation from projections, for example due to weather, unanticipated new sales, or cancelled contracts, the budget will be adjusted, generally on a quarterly basis. The importance of monitoring the budget on a daily basis is emphasized by the following statement,

Real-time budgeting improves reaction time and enhances profitability[1]

The monthly distribution of projected expenses and revenue is made on an income statement. Incorporating actual figures adjacent to the budgeted projections provides instant awareness of the current financial status of the company.

The projected income statement establishes a **cost structure** that reflects what percentages of revenues will be accounted for by costs. These percentages become icons within a budget for managing costs. One of the most critical of the structure components is direct costs. This component will determine whether the projected gross margin percentage is attained. The **gross margin** is the amount of money that is left after direct costs are paid. If the projected gross margin percentage (46%) is not attained due to over-budget direct costs, then the bottom-line profit projection (10%) will not be met. Since indirect costs ($264,000) and overhead costs ($924,000) are deducted from the gross margin dollars, a reduced amount directly impacts net profit dollars. To ensure that this gross margin goal is met, management places high priority on the management of the direct costs.

Table 2.13 BradCo Landscape Company March Budget Report

ENHANCEMENT DIVISION MARCH 2017

	Budget		Actual		Variance	
Revenue	$206,250	100%	$206,250	100%	0	
Direct costs	111,375	54%	113,348	55%	1,973	1%
Gross margins	94,875	46%	92,902	45%	1,975	1%
Indirect costs	16,500	8%	16,500	8%	0	
Overhead expenses	57,750	28%	57,750	28%	0	
Net pretax profit	**20,625**	**10%**	**18,652**	**9%**	**0**	

Budget development

Accounting departments post revenues and expenses as they are incurred and provide spreadsheets of the budget that show actual expenses and revenues adjacent to the projected figures. At a glance, managers can then ask why specific costs are over budget. If the cost is labor, then a weekly job cost analysis will be done to detect where the additional hours have been expended.

The monthly distribution of projected revenues and expenses actually establishes monthly income statements. Based on the seasonal patterns of the landscape industry, monthly figures will reflect a profit or loss. Maintenance companies that prorate annual contracts over a twelve-month period (e.g., $60,000 contract, $5,000/month) find that their costs will exceed revenue during the peak season and therefore show a loss for several months. Since these companies receive the same revenue flow in the off season, November–March, when labor costs (in regard to servicing the contracts) are lower, they will actually show a profit in the off season.

A monthly budget report enables management to immediately see variances. The red flags raised by the over-budget direct costs and indirect costs alerts management to production problems with one or more of its profit centers. Once the source(s) of the variances is determined, an analysis of job cost reports will be made. In this example, management may detect an excess of hours expended on one or more maintenance services, such as mulching or pruning associated with specific accounts. Further analysis may indicate that excess travel time due to road construction contributed to the increase in production hours.

The company's operating budget is delineated into division budgets as indicated in Table 2.14. The manager of BradCo's enhancement division will compare actual cost and revenue figures against the projected budget and proactively make the necessary adjustments. The actual budget figures are a direct reflection of the manager's abilities.

The monthly distribution of projected revenues and expenses creates monthly income statements. Based on the seasonal patterns of the landscape industry, monthly figures will reflect a profit or loss. This is illustrated in the six-month budget for BradCo Landscape Company's installation division.

Maintenance companies bill their clients on a monthly basis, prorating the total dollar amount of the contracts over a twelve-month period. In this monthly budget, there are several months in which costs exceed revenue, reflecting losses. This occurs during the peak season when labor hours are high due to the delivered services. Since the revenue is constant over the twelve-month period, profitable months for the maintenance sector are in the off season (November–March), when labor costs (in regard to servicing the contracts) are lower.

In contrast, installation/construction companies have budgets for each project and monitor actual expenses against the budgeted expenses. The project managers will monitor these expenses daily and make adjustments wherever possible to meet these projections.

Daily and weekly budget reports enable a manager to immediately see where the budget is off track. Then adjustments can be made to offset a decrease in gross margin. There may

Budget development

Table 2.14 2017 Monthly Budget Projections for BradCo Landscaping Company Enhancement Division

ENHANCEMENT DIVISION

	Annual	January	February	March	April	May	June
Revenue	$625,000	$11,875	$9,375	$22,750	$47,062	$97,185	$90,375
DIRECT COSTS							
Plants	$156,655	$1,550	$1,350	$4,471	$13,052	$26,100	$18,228
Hardscape	45,032	967	1,150	2,712	6,059	8,022	8,618
Labor	92,356	2,506	2,337	6,490	10,825	11,350	11,043
Labor burden	18,471	512	467	1,181	2,165	22,650	2,210
Subcont.	20,698	0	0	0	2,250	3,125	5,750
Total	$332,212	$5,535	$5,304	$14,854	$34,351	$71,247	$45,849
Gross margin	$292,788	$6,340	$4,071	$7,896	$12,711	$25,938	$44,526
OVERHEAD COSTS							
Indirect costs	$48,079	$4,582	$3,562	$3,688	$5,813	$4,850	$4,281
Equipment costs	$20,180	$1,193	$1,225	$1,451	$2,169	$2,225	$2,539
Administration	$154,993	$13,821	$12,912	$11,512	$14,188	$11,962	$11,061
Total overhead	$223,252	$19,596	$17,699	$16,651	$22,100	$18,797	$20,881
Net profit	$69,536	($13,256)	($13,628)	($8,755)	($10,611)	$7,141	$23,375

have been extenuating circumstances, such as equipment breakdowns that resulted in higher labor costs from overtime charges. In this situation, the manager may decide to allocate additional equipment and personnel to complete the job in a shorter period of time, thereby bringing down the unit cost of labor. If the equipment cost is over budget, it may be because the equipment is being used for only two hours but is being charged against the job for eight hours. Scheduling the equipment with other managers would result in more efficient use of the equipment and a reduction in overhead.

An operating budget is a financial tool that management must skillfully employ to enhance bottom-line profit.

SUMMARY

What is the budget development process?

It is a process that projects a company's revenues and expenses over a specific period of time. Operating budgets are developed for the overall company and further delineated into division budgets.

Budget development

The development process engages management in forecasting revenues, personnel, working capital, and fixed assets; determining profit requirements; establishing a cost structure as a percentage of revenues; and projecting overhead costs.

Why is profit a starting point for budget development?

Profit is the starting point because it is required to retire debt, provide excess compensation for personnel, purchase fixed assets, and provide capital for implementation of strategic plan goals.

How are budgets used as financial tools?

Budgets reflect the current financial status of a company or division by relating actual expenses and revenues to corresponding budgeted projections. Managers use these comparisons to detect variances and determine factors contributing to the variances.

How does budget management contribute to financial stability?

Budgets are a road map to a company's profitability. Budget management ensures that budget projections are met and that profitability is enhanced. Adjustments are made when necessary to keep the budget on track toward the profit goal.

How are variances in budget reports interpreted?

Managers access daily reports that indicate variances from budgeted projections. Interpretation involves investigating factors contributing to the respective budget line and determining where adjustments can be made.

KNOWLEDGE APPLICATION

1. Develop an annual budget based on your anticipated salary and living expenses. Distribute your expenses (rent, automobile insurance, entertainment, and so forth) and income on a monthly basis based on when they occur.

Budget development

2. University Landscape Company, with nine FTEs and sales of $500,000 based on sales of $55,000/employee, projects an increase in sales next year of $200,000. How many FTEs need to be recruited to meet this projection? The company has a 30% employee turnover rate.
3. You are a profit center manager who is in the process of developing your annual budget. The company president has offered managers a bonus incentive of $1,000 for every 1% increase they achieve in their gross margin. How would you manage your budget to be eligible for the bonus incentive?
4. Landscape company A, with gross sales of $700,000, has an annual net profit of 6% while landscape company B, with the same sales volume, has a net profit of 10%. Illustrate in an income statement format the net profits as stated. Explain your cost allocations in regard to their influence on the respective net profits.

NOTE

1. Charles Bowers, 2004, CEO Garden Gate Landscaping Inc., Personal Communication.

3

Profitable pricing

CHAPTER OBJECTIVES

To gain an understanding of:

1. Pricing components and their significance to price determination
   a. Direct costs
   b. Overhead costs
   c. Profit
2. The pricing formula's application to the landscape industry

   Direct costs + overhead costs + profit = selling price

3. Overhead markup and how it is calculated for labor and materials

   $$\text{Materials overhead markup} = \frac{\text{Overhead cost}}{\text{Materials costs}}$$

   $$\text{Labor overhead markup} = \frac{\text{Overhead cost}}{(\text{Direct labor costs} + \text{Labor burden costs})}$$

4. The dual overhead recovery method (DORM), its calculation and application

   Total DORM overhead = Materials overhead markup + Labor overhead markup

   $$\text{Materials overhead markup} = \frac{\text{Overhead cost}}{(X)(\text{Labor} + \text{Labor burden costs}) + \text{Material costs}}$$

   $$\text{Labor overhead markup} = \frac{(X)(\text{Overhead cost})}{(X)(\text{Labor} + \text{Labor burden costs}) + \text{Material costs}}$$

5. Hourly rate determination for labor and equipment

   Equipment cost per hour = Acquisition cost per hour + Maintenance cost per hour + Fuel cost per hour

Profitable pricing

KEY TERMS

acquisition cost per hour (ACPH)	indirect labor	materials to labor ratio (M/L)
breakeven point (BE)	inverse factor (INV)	multiple overhead recovery system (MORS)
complement markup	labor burden (LB)	overhead and profit per hour (OPPH)
direct costs	labor hourly rate selling price	
dual overhead rate method (DORM)	labor overhead markup (LOM)	overhead costs
	lifetime hours	overhead markup (OHM)
equipment cost per hour	maintenance cost per hour (MCPH)	overhead weighting factor
final selling price	material costs	profit
fuel cost per hour (FCPH)	material markup	profit markup (PM)
hourly labor rate	material overhead	

The title of this chapter appears to be stating the obvious. After all, why would any business not price its product profitably? It wouldn't do so intentionally, but if its management lacks an understanding of price components, it can very easily perpetuate unprofitable pricing. These components are direct costs, overhead, and profit, which collectively contribute to the pricing process.

Direct cost + overhead costs + profit = selling price

Direct Costs	Overhead Costs	Profit
Labor costs	Administrative salaries	Growth capital
Materials	Travel	Debt retirement
Equipment	Maintenance shop	Fixed asset purchases
Subcontracts	Sales department	Bonuses
	Equipment maintenance	Income tax
	Advertising	
	Professional development	

Direct costs are all of the expenses associated with the landscape service, such as labor, materials, subcontractors, and equipment rentals. **Overhead costs** are all the expenses incurred in running the business, including liability insurance, administrative/managerial salaries, vehicles and travel expenses, communication, equipment maintenance, sales and advertising, and professional development. **Profit** is the money remaining after all financial liabilities have been paid, which is used to retire debt, purchase fixed assets, provide employee bonus compensation, pay income tax, and invest in company growth.

The pricing process utilizes a company's financial data to provide an objective basis for determining a selling price. Without this process, pricing is – at best – an educated guess, particularly when it entails trying to meet a competitor's low price. This is a dangerous road that

Profitable pricing

often leads to a dead end or, in business terms, a bankruptcy. Remember those lumberyards and hardware stores that tried to compete on price with Home Depot? Where are they today? The same place as those electronics stores that tried to compete with Best Buy. But in spite of the presence of food warehouses and Wal-Mart, there is still a plethora of supermarkets. How have they survived? Their survival can be attributed to their management's pricing structure, which accounts for costs and profit. That is precisely what this chapter emphasizes: pricing for profit.

MATERIAL MARKUP

Pricing is not as simple as assigning a dollar figure to a product or service. When a retail nursery prices a 2.5" caliper tree, the determination is based on the wholesale cost (direct cost) plus a percentage over that cost to recover overhead expenses. This percentage increase is referred to as an **overhead markup (OHM)**. The overhead markup is calculated by dividing the **overhead costs** (O) by the **material costs** (M), derived from the nursery's income statement. Once the nursery adds the overhead markup to the material cost, the price represents a **breakeven point (BE)** at which the nursery will recover the direct cost of the tree and the overhead expenses associated with its sale. The calculation of the final selling price involves a **profit markup (PM)**, which is added to the breakeven point to attain the desired profit.

The nursery example illustrates an application of the overhead markup process to a typical retail business whose income revenue is based solely on the sale of merchandise or materials. All of the overhead expenses (sales personnel, rent, insurance, vehicles, and so forth) of such businesses support the sale of the merchandise materials, so the relationship between overhead expenses and material costs is used to calculate the overhead markup:

$$\text{Overhead markup (OHM)} = \frac{O}{M}$$

Using the numbers from the income statement in Table 3.1:

Table 3.1 Terrapin Nursery 2018 Income Statement

Terrapin Nursery

Total revenue	$600,000	100%
Material costs (M)	$330,000	55%
Total direct costs:	**$330,000**	**55%**
Gross margin: (600,000 − 330,000)	**$270,000**	**45%**
Overhead costs (O)	$210,000	35%
Pre-tax profit margin (PM) (270,000 − 210,000)	$60,000	10%

Profitable pricing

Overhead costs (O):	$210,000
Material costs (M):	$330,000
Overhead markup (OHM) $= \dfrac{O}{M} = \dfrac{210{,}000}{330{,}000} =$	.6364

The overhead markup (.6364, or 63.64%) is calculated and added to the material costs to yield the breakeven point that allows the nursery to recover the ongoing expenses of operating the retail business:

Breakeven point (BE) = M + (M × OHM)

Using the cost of a single tree as an example:

Tree cost (M):	$150.00
Overhead markup amount (M × OHM): (150 × .6364)	95.46
Breakeven point (BE) = M + Overhead markup amount:	$245.46

Using the profit percentage indicated in the income statement (Table 3.1), the final calculation involves a profit markup (PM) on the breakeven point (Table 3.2). The dollar amount of profit markup can be calculated from the breakeven point. It is then added to the breakeven point to yield the **final selling price**.

Table 3.2 Nursery Price Determination

Desired profit Margin (PM)	(100 − 10)/100 = Inverse factor (INV)	PM/INV = Complement markup (CM)
10% (.10)	.90	$\dfrac{.10}{.90} = .1111$
	Breakeven point (BE)	$245.46
	Profit markup amount (245.46 × .1111) =	27.27
	Final selling price (245.46 + 27.27) =	$272.73

To summarize:

Direct cost of 2.5″ caliper tree:	$150.00
+	
Overhead markup amount (OHM): ($150 × .6364)	95.46
Breakeven point (BE):	$245.46
+	
Profit markup amount (PM): ($245.46 × .1111)	27.27
Final selling price:	$272.73

Profitable pricing

To find the dollar amount of profit markup, first subtract the desired profit margin (PM) from 100 and then divide by 100 to find the **inverse factor (INV)**, expressed in decimals. Divide the desired profit margin (.10) by the inverse factor (.90) to determine the **complement markup (CM)**. Multiply the breakeven point by this number and add the resulting dollar amount to the breakeven point as a profit markup to yield the final selling price:

Continuing with the tree example for Terrapin Nursery:

The table below lists the formulas to determine profit markup amount. A range of profit margins is listed in column A, with corresponding inverse factors and complement factors in columns C and D. The example breakeven cost of $245.46 has been entered in all cases to compare the wide variety of results in columns F (profit markup) and G (final selling price).

$$\text{Final selling price} = \frac{\text{BE}}{\text{INV}}$$
$$= \frac{245.46}{.90}$$
$$= \$272.73$$

This can be explained by revisiting Table 3.2 and assigning percentage values to its components. If the final selling price represents 100% and the profit markup is a percentage of that amount, this means the breakeven point corresponds to 100% minus the profit markup percentage, which in this case yields 90%, or .90:

Direct cost of 2.5" caliper tree	$150.00		
+			
Overhead markup (OHM): ($150 × .6364)	95.46		
Breakeven point (BE):	$245.46	(.90)	90%
+			
Profit markup (PM):		(.10)	10%
Final selling price:	?	(1.00)	100%

The final selling price can also be calculated by simply dividing the breakeven point by the inverse factor of the desired profit margin:

If $245.46 is 90% of the final selling price, what would 100% of the final selling price be? The following proportion represents the relationship:

$$\frac{245.46}{.90} = \frac{\text{Final selling price}}{1.00}$$

Profitable pricing

Table 3.3 Profit Markup Factors

A	B	C	D	E	F	G
Desired Profit Margin (PM) (percentage format)	Desired Profit Margin (PM) (number format)	Inverse Factor (INV) = 1 − B	Complement Factor (CF) = B/C	Breakeven Point (BE) (currency format)	Profit Markup (PM) = E × D	Final Selling Price = E + F
1%	.01	.99	.0101	$245.46	$2.48	$247.94
2%	.02	.98	.0204	$245.46	$5.01	$250.47
3%	.03	.97	.0309	$245.46	$7.59	$253.05
4%	.04	.96	.0417	$245.46	$10.23	$255.69
5%	.05	.95	.0526	$245.46	$12.92	$258.38
6%	.06	.94	.0638	$245.46	$15.67	$261.13
7%	.07	.93	.0753	$245.46	$18.48	$263.94
8%	.08	.92	.0870	$245.46	$21.34	$266.80
9%	.09	.91	.0989	$245.46	$24.28	$269.74
10%	.10	.90	.1111	$245.46	$27.27	$272.73
11%	.11	.89	.1236	$245.46	$30.34	$275.80
12%	.12	.88	.1364	$245.46	$33.47	$278.93
13%	.13	.87	.1494	$245.46	$36.68	$282.14
14%	.14	.86	.1628	$245.46	$39.96	$285.42
15%	.15	.85	.1765	$245.46	$43.32	$288.78
16%	.16	.84	.1905	$245.46	$46.75	$292.21
17%	.17	.83	.2048	$245.46	$50.27	$295.73
18%	.18	.82	.2195	$245.46	$53.88	$299.34
19%	.19	.81	.2346	$245.46	$57.58	$303.04
20%	.20	.80	.2500	$245.46	$61.37	$306.83
25%	.25	.75	.3333	$245.46	$81.82	$327.28
30%	.30	.70	.4286	$245.46	$105.20	$350.66
40%	.40	.60	.6667	$245.46	$163.64	$409.10
50%	.50	.50	1.0000	$245.46	$245.46	$490.92

Source: Adapted from *Pricing for the Green Industry*, 3rd ed., Frank H. Ross, Ross-Payne & Associates. © 2012, Professional Landcare Network (renamed National Association of Landscape Professionals), Herndon, VA.

Simply cross-multiply the equation to find the final selling price:

$$(245.46)(1) = (.90)(\text{Final selling price})$$
$$\frac{245.46}{.90} = \text{Final selling price}$$
$$272.73 = \text{Final selling price}$$

Profitable pricing

Why doesn't it work to calculate the profit markup by multiplying the breakeven point by .10 ? The reason is illustrated by the following example:

Material cost of 2.5" caliper tree	$150.00
Overhead markup: (150 × .6364)	95.46
Breakeven point:	$245.46
Profit markup (miscalculation): (245.46 × .10)	24.55
Final selling price:	$270.01

Calculated this way, the profit markup results in a 9% ($24.55/$270.01) markup rather than the desired 10% reflected in the correct final selling price of $272.73. This is because the profit markup is a percentage of the *final selling price*, not the breakeven point. The $2.72 difference in price might seem insignificant for an individual tree sale, but it becomes significant when this 9% markup is applied to the sale of multiple trees. Five hundred 2.5" caliper trees priced with a miscalculated 9% markup would generate $1,360 less in income revenue than those priced with the desired 10% profit.

It is apparent in the previous pricing example that a profit markup miscalculation of even 1% can have a major negative impact on a company's profitability. The cumulative effect of such small but consistent miscalculations results in a smaller profit margin, compromising a company's ability to reinvest in the company's future growth.

The pricing process becomes more complex for a landscape contracting business whose revenue is generated by the sale of labor, materials, equipment, and subcontracts. In addition, landscape companies have multiple divisions, such as maintenance, installation, enhancement, construction, and irrigation, all of which have their own respective labor and equipment costs and varying overhead rates. Therefore, *a landscape company must have a comprehensive financial database for each division to price its jobs profitably*. Keeping the database up to date is imperative to account for increases in direct and overhead costs, such as material costs, insurance rates, fuel costs, and hourly labor rates.

A landscape company's calculation of an overhead markup is contingent upon its primary source of revenue. In contrast to the previous nursery example in which the overhead markup was based on material sales, landscape companies recover their overhead with a labor markup or a combination markup on both labor and materials, called a *dual overhead rate*, which is discussed in more detail later on. The proportion of labor and material costs associated with a particular job will determine which markup method is used.

LABOR MARKUP

An income statement profile of a landscape maintenance company is illustrated in Table 3.4.

Profitable pricing

Table 3.4 BradCo Landscape Company 2017 Income Statement

Total revenue:	$1,500,000	100%
Direct job costs		
Direct labor (L)	$450,000	30%
Labor burden (LB)	$45,000	3%
Material	$225,000	15%
Subcontractors	$75,000	5%
Total direct costs:	**$795,000**	**53%**
Gross margin: (1,500,000 − 795,000)	**$705,000**	**47%**
Overhead costs (O)	$600,000	40%
Pre-tax profit margin (PM)	$105,000	7%

With labor as the predominant direct cost factor, the major portion of overhead costs are all labor related. The formula for calculating the labor overhead markup is:

$$\text{Labor overhead markup (LOM)} = \frac{O}{L+LB}$$

Using the numbers in the table above:

Overhead costs (O):	$600,000
Direct labor (L):	$450,000
Labor burden (LB):	$45,000

$$LOM = \frac{O}{L+LB} = \frac{600{,}000}{450{,}000+45{,}000}$$
$$= \frac{600{,}000}{495{,}000}$$
$$= 1.21 \text{ or } 121\%$$

Determining the company's hourly selling price per laborer (Table 3.5) begins with adding **labor burden (LB)** – the costs associated with state and federal payroll taxes and worker's compensation insurance – to the base hourly rate, then applying the labor overhead markup, and finally adding the desired profit markup. The breakeven hourly rate of $35.78 allows BradCo to recover their labor overhead costs. Included in the **labor overhead markup (LOM)** is uniform rental, health insurance, general liability insurance, hand tools, equipment maintenance, sick day compensation, holiday and annual vacation leave compensation, and in

Profitable pricing

Table 3.5 BradCo Landscape Company Labor Hourly Rate Calculation

Laborer hourly rate (L):	**$12.65**
Labor burden (LB): (28%)*	3.54
Labor cost (L + LB):	$16.19
Labor overhead markup (16.19 × 121%)	+ 19.59
Breakeven hourly rate: (93%)	**$35.78**
Pre-tax profit margin (7%)	+ 2.69 (35.78 × .0753)
Labor hourly rate selling price (100%)	**$38.47**

* varies with state taxes

some cases pension plan contributions, as well as sales department salaries and expenses, professional development, and administrative salaries and expenses. Based on an hourly rate of $12.65 and adding the appropriate labor burden, a labor overhead markup, and a profit markup of 7%, the **labor hourly rate selling price** is calculated to be $38.47. It is evident from this calculation that overhead has a significant impact on the hourly labor selling price. A company's selling price can therefore be much more competitive if it is able to reduce its overhead expenses.

$$\text{Labor hourly selling price} = (L+LB) + LOM + PM$$

DUAL OVERHEAD RATE METHOD

Overhead recovery calculated individually for labor and materials is appropriate if the sales of a landscape company are dominated by either one or the other. But what about design/build companies and landscape construction companies that have a significant amount of sales both in materials and in labor? This question was posed to a think tank at the Fails Management Institute in the 1970s. Dr. Fails, Frank Ross, and some of their colleagues at North Carolina State University pondered the question posed by the landscape industry and ultimately came up with a solution. That solution was the **dual overhead rate method (DORM)**, which calculates overhead recovery based on the ratio of material to labor costs.

The calculation of a dual overhead rate begins with establishing a materials to labor ratio based on direct costs projected in the company's income statement. An example of an abbreviated income statement projection for a design/build landscape business is presented in Table 3.6.

Profitable pricing

Table 3.6   Potomac Landscape Company 2017 Income Statement

Total revenue:	$3,000,000	100%
Direct costs		
Materials (M)	$900,000	30%
Labor plus labor burden (L)	$600,000	20%
Subcontracts	$150,000	5%
Other	$30,000	1%
Total direct costs:	**$1,680,000**	**56%**
Gross margin:	**$1,320,000**	**44%**
Overhead costs (O)	$1,050,000	35%
Pre-tax profit margin (PM)	$270,000	9%

The **materials to labor ratio (M/L)** is calculated by dividing the total material direct costs by the total of direct labor plus labor burden costs:

$$\text{Materials to labor ratio} = \frac{M}{L}$$

Plugging in the numbers from the table above:

$$\frac{M}{L} = \frac{900{,}000}{600{,}000} = 1.5$$

This ratio is indicative of the proportion of Potomac Landscape's project expenditures allocated to materials and labor. Based on this projection, Potomac can anticipate that for every dollar of production labor there will be a corresponding expenditure of $1.50 for materials.

Determination of this ratio provides a basis for the dual overhead rate method to allocate overhead costs proportionally for those expenses that support materials versus labor. As the M/L ratio increases, the DORM allocates a greater proportion of the overhead costs to materials, and as it decreases, the allocation amount is increased on the labor side.

The next step in the dual overhead rate calculation requires an **overhead weighting factor**. This factor is obtained by referring to the chart of overhead weighting factors (developed by Dr. Emol Fails) in Table 3.7. Potomac Landscape's M/L ratio of 1.5 has an overhead weighting factor of 2.79. If an M/L ratio falls between two factors, round it to the nearest M/L ratio. The weighting factor (X) is the key component in the dual overhead rate formula that determines the distribution of overhead to materials and labor.

Profitable pricing

Table 3.7 Overhead Weighting Factors

M/L – Materials to labor ratio
X – Overhead weighting factor

M/L	X	M/L	X	M/L	X	M/L	X
.00	1.90	1.00	2.46	2.00	3.13	4.00	4.42
.05	1.84	1.05	2.49	2.10	3.19	4.20	4.54
.10	1.88	1.10	2.53	2.20	3.26	4.40	4.66
.15	1.91	1.15	2.56	2.30	3.33	4.60	4.77
.20	1.95	1.20	2.59	2.40	3.39	4.80	4.88
.25	1.98	1.25	2.62	2.50	3.46	5.00	5.00
.30	2.01	1.30	2.66	2.60	3.53	5.20	5.10
.35	2.04	1.35	2.69	2.70	3.59	5.40	5.21
.40	2.07	1.40	2.72	2.80	3.66	5.60	5.31
.45	2.10	1.45	2.75	2.90	3.72	5.80	5.41
.50	2.13	1.50	2.79	3.00	3.79	6.00	5.51
.55	2.16	1.55	2.82	3.10	3.85	6.30	5.65
.60	2.20	1.60	2.86	3.20	3.92	6.70	5.83
.65	2.23	1.65	2.90	3.30	3.98	7.00	5.96
.70	2.26	1.70	2.93	3.40	4.05	7.50	6.15
.75	2.30	1.75	2.96	3.50	4.11	8.00	6.35
.80	2.33	1.80	2.99	3.60	4.17	8.50	6.52
.85	2.36	1.85	3.03	3.70	4.23	9.00	6.68
.90	2.39	1.90	3.06	3.80	4.30	9.50	6.82
.95	2.43	1.95	3.09	3.90	4.36	10.00	6.95

Source: Adapted from *Pricing for the Green Industry*, 3rd ed. Frank H. Ross, Ross-Payne & Associates. © 2012, Professional Landcare Network (renamed National Association of Landscape Professionals), Herndon, VA.

Dual rate formulas

The overhead weighting factor (X), labor and labor burden costs, material costs, and overhead costs are the components of the formulas for calculating the markup on materials and the markup on labor.

Potomac Landscape Company

Overhead weighting factor (X): (for this example) 2.79
Labor and labor burden costs (L): $600,000
Material costs (M): $900,000
Overhead costs (O): $1,050,000
Materials markup = O ÷ (X)(L+LB) + M
Labor markup = (X)(O) ÷ (X)(L+LB) + M

Using the numbers in the example above:

Material markup = $\dfrac{1{,}050{,}000}{(2.79)(600{,}000) + 900{,}000}$

= $\dfrac{1{,}050{,}000}{2{,}574{,}000}$

= 41%

Using the numbers in the example above:

Labor markup = $\dfrac{(2.79)(1{,}050{,}000)}{(2.79)(600{,}000) + 900{,}000}$

= $\dfrac{2{,}929{,}500}{2{,}574{,}000}$

= 114%

In this example, the labor and labor burden are weighted an additional 2.79 times over the material costs. The basis for application of the additional weight factor to labor and labor burden is to account for the greater portion of overhead costs associated with labor versus materials. Some of the larger overhead costs that support labor include:

Indirect labor – non-billable labor compensation, e.g., vacation, sick leave, downtime associated with loading and unloading, picking up material, equipment breakdowns, and so forth.
Payroll taxes – federal, state, local, social security, workers' compensation
Communications – cell phones, tablets
Liability insurance – vehicles and equipment
Sales expenses – salaries and salary burden
Administrative expenses – payroll, human resources, professional development

The following calculation validates that the calculated markups will recover all of the overhead costs that are applied to the budgeted costs for labor and materials:

Labor overhead	= L × Labor markup
	= $600,000 × 114%
	= $684,000
Material overhead	= M × Material markup
	= $900,000 × 41%
	= $369,000
Total overhead	= Labor overhead + Material overhead
	= $684,000 + $369,000
	= $1,053,000

Profitable pricing

The management of the Potomac Landscape Company would be very satisfied with the results of the dual overhead rate markups, since the variance is on the plus side, recovering $3,000 over the projected annual overhead of $1,050,000.

The following summarizes the application of the three markups methods discussed:

Material markup – used by companies whose sales are primarily materials, e.g., retail and wholesale nurseries.

Labor markup – used by landscape companies or divisions whose sales are dominated by labor-intensive services, e.g., maintenance, snow plowing, pressure washing, and so forth.

Dual rate overhead method – applicable to installation jobs that have variable amounts of materials and labor.

Design/build companies and installation divisions are examples of companies for which the DORM would be effective in overhead recovery.

OTHER OVERHEAD RECOVERY METHODS

The above methods are just a few examples of how overhead recovery can be calculated. There are several other methods that are used in the landscape industry, such as **MORS (multiple overhead recovery system)** and **OPPH (overhead and profit per hour)**. Detailed discussions of these and other methods are found in *How to Price Landscape and Irrigation Projects*, by James Huston (J.R. Huston Consulting, Inc.). All are built on the premise of having accurate financial information and budget projections. The key to overhead recovery is calculating the percentage of overhead markup based upon the ratio of material to labor costs.

Companies that have been in business for several years have a financial database that provides a documented history of cost percentages and overhead costs for individual jobs and collectively for their divisions. These percentages may be consistent from year to year if the company is engaged in similar contracts (for example, maintenance or irrigation), or the percentages may vary if the company is engaged in bidding commercial installation contracts. The latter is much more variable due to site conditions, installation components, wage scales, and subcontractors.

The most reliable overhead recovery system is one that is developed internally with a company's financial database as its foundation. A company may use a MORS, DORM, or OPPH as the infrastructure of its system with modifications that address specific job situations, such as amount of labor, labor burden, equipment, and materials. Just as with

Profitable pricing

accounting software programs and men's suits, overhead recovery systems are rarely perfect fits off the rack.

EQUIPMENT PRICING STRUCTURE

The previous pricing discussion dealt with recovering overhead costs associated with the direct costs of materials and labor. This was achieved by applying markup methods on the basis of the amount of materials or labor that were associated with the revenues generated.

Another important overhead expense recovery method pertains to the equipment employed by landscape companies. The landscape industry, in addition to being labor intensive, is also equipment intensive. Equipment ranging from lawn mowers to backhoes and trenchers comprise a significant portion of a company's fixed assets and overhead costs. Associated with such equipment is the initial capital investment, as well as maintenance costs. In order to recover these costs, a pricing structure needs to be established for each piece of motorized equipment, including trucks. Depending on how much equipment is used on a job, landscape companies categorize it as either a direct cost or an overhead expense. The following examples illustrate the process involved in establishing hourly equipment rates.

The primary cost factor components for pricing equipment at hourly rates are:

1 **Acquisition cost per hour (ACPH):**
   a   Purchase price (PP) – including taxes and registration fees
   b   Loan interest (LI) – over the duration of the loan
   c   Trade-in value (TV)
   d   Lifetime hours (LH)
2 Maintenance cost per hour (lifetime projected costs)
3 Fuel cost per hour

Trucks are collectively the most costly equipment items among a landscape company's fixed assets. They are used by supervisors and crew leaders to transport crews and materials as well as by account managers whose daily travel may entail site estimates, inspections, contract renewals, job supervision, and sales presentations. Table 3.8 calculates an hourly cost that would be charged for every hour that a crew cab pickup truck is on a job.

Assume that a truck will be used for four years and then traded in. The lifetime hours can be calculated using the following formula:

    Lifetime hours (LH): 40 hours $\times$ 52 weeks $\times$ 4 years = 8,320

Profitable pricing

Table 3.8 Equipment Pricing: Crew Cab Pickup Truck

1. **Acquisition cost per hour (ACPH)**
   Purchase price (PP): $62,000 (inclusive of taxes and registration fees)
   Loan interest (LI): Purchase price × years of payments × interest rate [one-half of loan term (2 years), based on equal monthly payments leaving an outstanding principal loan balance for one-half the period]
   $62,000 × 4 × .02/2 = $2,480
   Trade-in value (TV): $20,000
   ACPH = (PP + LI − TV)/LH
   ACPH = ($62,000 + 2,480−20,000)/8,320 hr = $5.35/hr

   ACPH = $5.35/hr

2. **Maintenance cost per hour (MCPH)**

License fees (4 years × $150)	$600
Insurance ($1,500/year × 4 years)	$6,000
Lube, oil, filters (20 × $40)	$800
Brake service and replacements (3 × $400)	$1,200
Clutch service and replacement (2 × $500)	$1,000
Smog certification (2 × $60)	$120
Tires (2 sets × $600)	$1,200
Miscellaneous maintenance (batteries, etc.)	$800
Engine replacement	<u>$1,167</u>
Total (MC):	$12,887

   MCPH = MC/LH
   MCPH = $12,887/8,320 hours = $1.55/hr

   MCPH = $1.55/hr

3. **Fuel cost per hour (FCPH)**
   (Fuel cost per gallon/miles per gallon) × (miles driven per day/8 hours)
   FCPH = (CPG÷MPG) × (M/H)
   FCPH = ($2.50/15 mpg) × (75 miles/8 hours) FCPH = $.1666 × 9.375 = $1.56

   FCPH = 1.56

4. **Crew cab truck total cost per hour = ACPH + MCPH + FCPH**
   $5.35 + $1.55 + $1.56 = $8.46

   Equipment cost per hour: $8.46

As is evident from the calculation in Table 3.8, the hourly rate recovers a portion of all the costs associated with the vehicle. Contrasting the application of this cost recovery method to a mileage assessment method illustrates the importance of its implementation.

Mileage Assessment Method

Job Mileage	Mileage Rate	Duration
30 miles	$.56/mile (IRS Allowance)	8 hours

Mileage cost assessment = $.56 × 30 mile = $16.80
Hourly rate assessment = $8.46 × 8 hr = $67.68

The above example shows that the mileage assessment method prescribed by the IRS is inadequate to recover the costs associated with purchasing, maintaining, and operating this vehicle. The cost difference increases exponentially with the number of crew cab trucks and the total mileage logged on a daily basis. Rather than include the truck costs as a direct cost, many companies will calculate it into an equipment line under overhead costs.

Every piece of equipment has an hourly rate established by using the same pricing components, which is coded for each piece of equipment and incorporated into estimates for landscape contracts either as direct costs or overhead costs. Companies can obtain equipment production rates, lifetime hours, and maintenance costs from owner's manuals and industry publications, such as *The Complete Estimating Book with Labor and Equipment Production Times for the Green Industry*, Vander Kooi and Associates, Inc., Littleton, Colorado.

SUMMARY

The construction of a pricing structure requires the use of two financial tools that were introduced in the previous two chapters. Those tools are the chart of accounts, which delineates between costs associated with what is sold (direct costs), the costs associated with supporting the sale (overhead), and the budget, which projects those costs as a percentage of annual sales.

What are the pricing components and what is their significance to price determination?

The pricing components consist of:

- Direct costs – labor, labor burden, materials, equipment, subcontractors. These are the primary cost factors associated with job production. A pricing structure is built on the percentage of production revenues.
- Overhead costs – administrative salaries, equipment maintenance, sales salaries, advertising, uniform rentals, professional development. These and other costs that support production are included in pricing as a percentage of the direct cost expenses.
- Profit – capital for growth, retirement of debt, fixed asset purchases, excess compensation, and income tax liabilities. A percentage of profit based on budget projections is calculated into the selling price.

Profitable pricing

How does the pricing formula apply to the landscape industry?

The pricing formula, Direct costs + Overhead + Profit = Selling price, is applicable to all sectors of the landscape industry. Computation of the pricing components varies based on the mix of labor, materials, and equipment.

How is markup calculated for labor and materials?

Markups for labor and materials enable companies to recover overhead costs associated with these direct costs.

Labor markup is determined by:

- Determining annual costs for overhead, labor, and labor burden
- Dividing overhead costs by the sum of labor and labor burden costs

(overhead/labor + labor burden)

Material markup is determined by:

- Determining annual costs for overhead and materials
- Dividing overhead costs by material cost (overhead costs/material costs)

Using the dual overhead rate method (DORM), how is the dual overhead recovery markup calculated and where is it applicable?

The dual overhead rate method is applicable to installation jobs in which the ratio of materials to labor varies. It calculates overhead markup based on the ratio of materials to labor and a weighting factor that equates to the respective ratio.

The dual overhead recovery is calculated by:

- Determining annual costs for labor (plus labor burden)
- Determining annual costs for material
- Determining annual costs for overhead
    1. Calculating the ratio of materials to labor for a job: (materials/labor)
    2. Determining the weighting factor for the respective ratio (from table)
    3. Calculating the labor markup:

(weighting factor)(overhead)/[(weighting factor)(labor + labor burden) + materials]

4. Calculating the material markup:

    overhead/[(weighting factor)(labor + labor burden) + materials]

5. Adding the labor markup and the material markup to yield the total overhead cost

How are hourly rates determined for labor and equipment?

Hourly labor rates are determined by:

- Adding a labor burden as a percentage to the base hourly rate
- Adding a markup for labor overhead
- Adding a profit percentage

Equipment cost per hour is determined by:

- Calculating an acquisition cost per hour (ACPH)
- Calculating a maintenance cost per hour (MCPH)
- Calculating fuel cost per hour (FCPH)
- Adding the costs together: ACPH + MCPH + FCPH

KNOWLEDGE APPLICATION

1. Buffalo Grove Nursery has annual material costs of $550,000 and overhead costs of $350,000. BGN's projected net profit is 10%.

    a. Calculate the overhead markup for materials.
    b. Calculate the breakeven point for a $750 maple tree.
    c. Calculate the final selling price for the maple tree.

2. a. Give three landscape industry examples of dual rate overhead applications.
    b. Why would the DORM method be preferred over individual material and labor markup methods?

3. Based on the financial information that follows:
    a. Calculate the breakeven hourly labor price.
    b. Calculate the hourly rate selling price.

Profitable pricing

Overhead costs	$525,000
Labor and labor burden	$350,000
Crew base hourly rate	$8.50
Labor burden base percent	28%
Projected profit	10%

4 a Which of the pricing components do you feel has the greatest impact on a company's competitive pricing structure?

   b Give examples of how companies can reduce direct costs.

4

Estimating

CHAPTER OBJECTIVES

To gain an understanding of:

1   The estimating system
    a   Direct costs + overhead + profit = selling price
2   Estimating maintenance jobs
    a   Square footage takeoffs
    b   Services
    c   Frequency
    d   Production hours
    e   Material quantities
3   Estimating installation jobs
    a   Job specifications
    b   Site analysis
    c   Square footage takeoffs
    d   Plant quantities
    e   Equipment hourly rates
    f   Production standards
4   Interfacing the estimating system with job cost management

Estimating

KEY TERMS

assembly	digitizer	proposal
bid	estimate	site preparation
contingency clause	landscape estimators	takeoffs
crew average wage (CAW)	on-screen takeoffs (OST)	time and materials estimate

A lot of responsibility lies on the shoulders of estimators. The estimates that they generate provide the foundation for establishing project budgets, which project managers will use to develop production schedules based on the projected material costs and labor hours.

Regardless of the project manager's skills and the production efficiency of the crews, the bottom-line profit is contingent upon the accuracy of the estimated labor and materials costs. How do estimators ensure the accuracy of their estimates? By becoming familiar with all aspects of the project.

Estimating is a process involving job site analysis and projection of cost factors associated with job production, following current financial data from the following resources:

- Accounting system – a database for accessing labor and material costs
- Production standards – time required for landscape operations
- Budget – projected profit requirement, overhead, and direct cost percentages

By definition, an **estimate** is an approximation or an educated guess. The initial estimate is reviewed by management and adjustments are made before a **commercial bid** or residential job **proposal** is submitted. Bids presented for commercial maintenance and construction contracts are often competing with other contractor submittals for the same job with the same specifications.

There are two distinct contract situations: negotiable residential client contracts, and non-negotiable commercial client contracts that have been awarded to the landscape company as the winning bid to carry out the specifications of a job. A landscape contractor can negotiate with a residential client until they concur on the price for the services and materials to be provided, which is subsequently presented in the form of a written proposal. On the other hand, the competitive bidding process for commercial contracts eliminates the possibility of renegotiating the job price after the contract has been awarded.

ESTIMATOR'S RESPONSIBILITIES

Landscape estimators determine what it takes to complete the job. The estimating process includes a thorough site analysis alongside the job specifications, which leads to a comprehensive estimate of material and labor cost projections. The company's financial history and

Estimating

budget projections provide the estimator with the information required for the selling price formula:

Direct Costs	+	Overhead	+	Profit	=	Selling Price
Estimator input		**Budget projections**				
		Financial history				

A maintenance estimate would include the cost for mowing, seasonal color installation, pruning, fertilizing, and herbicide and pesticide applications, while a construction estimate might include costs for deck construction, retaining walls, water features, and plant installations, lighting, and irrigation.

In both maintenance and construction estimates, the estimator needs to determine square footage for the designated areas on the job site to make what is referred to as **takeoffs**, which is the list of materials needed for the job. The estimator "takes off" the types and quantities of items needed for the job by evaluating the site measurements. Linear measurements can be made manually with a tape measure or digitized measuring wheel on the job site and used to make a field sketch, or the client may provide a sketch of site plans or a set of blueprints. A **digitizer** is an electronic measuring device equipped with a board and a pointing tool that is used to digitize linear measurements from a blueprint to generate takeoffs. A maintenance estimate would have takeoffs pertaining to turf, shrub, flower bed areas, and tree rings. Landscape construction takeoffs determine areas associated with plant and sod installations, hardscape features such as patios and decks, retaining walls, quantities of plant materials, and irrigation system components, along with items associated with the installations.

The current trend is toward making **on-screen takeoffs (OST)** by using takeoff and estimating software. Using on-screen takeoff programs eliminates the need for blueprints and has proven to expedite the estimating process. However, it is still imperative to visit the job site to assess the topography, exposure, soil conditions, and existing plant materials and to establish the scope of maintenance operations.

To summarize, the key estimate components compiled by the estimator are:

1. **Direct costs** – all material costs, labor costs, subcontractors, and equipment.
2. **Indirect costs** – collateral costs associated with the job such as permits, bonds, equipment mobilization, supervision time, travel time, silt fences, and so forth.

The remaining portions of the estimate involve applying standardized calculations for recovering overhead, accounting for specific job conditions, and warranties:

3. **Markups** – calculated to recover overhead costs associated with labor and materials. A budgeted markup percentage is often used for maintenance estimates, in contrast

Estimating

to landscape construction estimates that may calculate markups on an individual job basis due to variations in the proportion of material to labor costs.

4   **Risk factor** – a percentage markup included for unanticipated conditions that may result in additional labor or materials expenses. Poor drainage or soil compaction on a construction site would be an example, or the discovery of unmarked utilities or old foundations and footings.
5   **Warranty** – a percentage markup of the total job cost, generally 3–5%, to cover the costs associated with replacement of materials under warranty.
6   **Sales tax** – on all materials used in the job (except for tax-exempt organizations and institutions).
7   **Profit** – based on the current budgeted amount to meet the company's financial needs. The dollar amount of profit is calculated as a percentage of the total job cost.

Profit percentages vary for landscape maintenance companies and construction companies depending on whether their client base is commercial or residential. A company may be willing to reduce its profit margin for a competitive commercial job in order to win a large bid or one with the potential for future jobs with the client or general contractor. The final estimate stipulates:

- What is to be done – job production
- How many labor hours – job completion
- Quantities of materials – per operation
- Total production costs
- Total selling price

PRODUCTION STANDARDS

In order to generate an estimate, a company uses its database of production standards, which is predetermined by a crew supervisor who tracks the production times of crews working on all phases of maintenance services, plant installation, or construction. These standards are adjusted when practices change due to the introduction of new equipment, such as motor-powered bed edgers, mulch blowers, or leaf vacuums. The information that is compiled will include:

- Area dimensions – square footage or linear feet
- Hours to completion (based on crew size and individual work hours)

Estimating

- Power equipment production time
- Material specifications and quantities

Once this database is established, an estimating software program can extrapolate the data into the respective production categories. Examples of production rates are given in Tables 4.1 and 4.2:

Table 4.1 Plant Installation Production Rates

Description	Unit	Qty / Work Hours*	Production Time	Equipment
Site preparation				
Grading to +/− .1/ft	SF	4–6,000	10–15 min/1,000 SF	tractor
Rototilling 6" depth	SF	1–2,000	40 min/1,000 SF	rototiller
Finish grade	SF	1,000	60 min/1,000 SF	
Planting (inclusive of placement to mulch ring preparation)				
trees (box diameter/balled & burlapped caliper)				
72" box	Ea	.125	8 hrs	1 hr backhoe
60" box/5" caliper	Ea	.167	6 hrs	.8 hr backhoe
48" box/4" caliper	Ea	.25	4 hrs	.5 hr backhoe
24" box/2" caliper	Ea	.50	2 hrs	.25 hr backhoe
Trees (containers)				
15 gallon	Ea	1.0	1 hr	
5 gallon	Ea	5.0	12 min (.2 hr)	
1 gallon	Ea	12–15	4–5 min (.06–.083 hr)	
Staking & guying	3 stakes	2	30 min (.5 hr)	
Shrubs (balled & burlapped)				
3'	Ea	.75	1.3 hrs	
2'	Ea	1.2	.8 hrs	
1–1/2'	Ea	2–2.5	.33–.5 hrs	
1'	Ea	3–4	15–20 min (.25–.33 hr)	
Shrubs (containers)				
15-gallon containers	Ea	1	1 hr	
5-gallon containers	Ea	4–5	12–15 min (.2–.25 hr)	
1-gallon containers	Ea	12–15	4–5 min (.083 hr)	
Ground cover				
2-1/4" pots	Flat	1.5–2	30–45 min/flat (.5–.75 hr)	
Sod	SF	225–300	20–27 min/100 SF (.33–.45 hr/100 SF)	

* Production of one worker within a one-hour period.
Source: *How to Price Landscape and Irrigation Projects*, Smith Huston, Inc., Englewood, CO

Estimating

Table 4.2 Maintenance Production Rates

Service	Equipment	Work Hours*	Comments
Mowing	21" rotary	1.75 hrs/10,000 SF	Mulching blade
		2.0 hrs/10,000 SF	Collecting clippings
	48" riding and 52"	1.0 hr/32,000 SF	Mulching blade
		1.0 hr/28,000 SF	Collecting clippings 1.6 mph
	60" mower	1.0 hr/1.25 acres	2.5–3.5 mph
Thatching	Walk behind	4.0 hrs/10,000 SF	
Fertilizing	Rotary spreader	1.0 hr/acre	Non-divided
		1.0 hr/20,000 SF	Residential property
	Drop spreader	1.0 hr/15,000 SF	
Herbicide application	Rotary spreader	1.0 hr/acre	Pre-emergent
	Sprayer	1.0 hr/15M SF	Post-emergent
Hedge trimming	Hedge	.25 hr/80 linear feet	5.6' × 5.6' hedge
		2.5 hrs/80 linear feet	Cleanup
Bed edging	Power stick edger	.35 hr/100 linear feet	Inclusive of cleanup
Tree ring edging	Hand	.35 hr/30" diameter	
		.50 hr/36–46" diameter	

* Production of one worker within a one hour period.

Source: *Guide to Growing a Successful Landscape Maintenance Business*, Professional Landcare Network, Herndon, VA, reprinted with permission

Once a company establishes its production rates, it has a database for estimating the total production hours for planting, mowing, and edging, and all other operations.

The formula for estimating total production hours is:

$$\text{Total production hours} = \text{Unit quantity} \times \text{Production standard} \times \text{Operation frequency}$$

Examples of production hours estimate for maintenance and installation operations follow. The figures for planting various sizes of shrubs can be found in Table 4.1. For the hedge trimming example, the production rate shown in Table 4.2 is .25 hr / 80 linear feet, with cleanup rated at 2.5 hr / 80 linear feet.

Planting ten 5-gallon shrubs:

Total production hours = 10 × (.25 hr/shrub) × 1
Total production hours = 10 × .25 hr × 1 = **2.5 hours**

The production hours estimate for planting ten 5-gallon shrubs, based on these standards, is two and a half hours.

Hedge trimming:

Total production hours = 160 ft × (.25 hr / 80 ft) × 7
Total production hours = 160 × .003 × 7 = **3.6 hours**

An additional consideration is the cleanup:

Total production hours = 160 × (2.5 hr / 80 ft) × 7
Total production hours = 160 × .03 × 7 = **33.6**

The total production hours estimate for trimming 160 ft. of hedge seven times is:

33.6 hr + 3.6 hr = **37.2 hours**

ESTIMATING PROCEDURE FOR MAINTENANCE CONTRACTS

The estimator assesses the site's landscape status in terms of overall conditions, topography, drainage, and soil conditions as well as the condition of the turf, shrub and seasonal color beds, and trees to determine the amount and type of work needed. This analysis will reveal whether renovation of any of the landscape entities is warranted: bed edging, tree rings, post-emergent weed applications, pesticide applications, turf aeration, liming and fertilization, corrective or rejuvenation pruning, drainage, grading, and so on. A renovation estimate is different from one presented for a year-round contract, since it includes the preliminary remedial work needed to elevate the maintenance standards of the property and begin the process of establishing a management program.

A conventional estimate for a maintenance contract would entail estimating services specified by the potential client, which may include mowing, trimming, edging, fertilizing, applying chemicals for weed, insect, and disease control, pruning, shearing, aerating and overseeding lawns, mulching, removing leaves, and in cold climates removing snow and deicing. An estimating form for such services is illustrated in Table 4.3.

Table 4.3 Landscape Maintenance Estimate Form

Client:		Location:			
Property Manager:		**Contract Period:**			
Telephone:		e-mail:			
Maintenance Services	Hrs./Service Frequency	Service Hours	Labor	Cost/Hour	Total cost
Turf					
Mowing, trimming, edging, blowing					
Fertilizing					
Herbicides					
Pre-emergent					
Post-emergent					
Insecticides					

(*Continued*)

Table 4.3 (Continued)

Shrubs					
Pruning					
Corrective					
Rejuvenation					
Shearing					
Fertilizing					
Herbicides					
Pre-emergent					
Post-emergent					
Insecticides					
Fungicides					
Mulching					
Flowerbeds					
Herbicides					
Fungicides					
Insecticides					
Fertilizing					
Mulching					
Trees					
Pruning					
Corrective (up to 5′)					
Fertilizing					
Insecticide					
Other					
Leaf removal					
Aeration, overseeding					
Perennial cutbacks					
Deicing					

Materials	*Quantity*	*Unit Cost*	*Total Cost*
Herbicides			
Pre-emergent			
Post-emergent			
Fungicides			
Insecticides			
Fertilizers			
Mulch			

Estimating

Materials	Quantity	Unit Cost	Total Cost
Other			
Surfactant			
Die marker			
Deicer			
Growth retardant			

Total labor cost _____

Total material cost _____

Labor markup _____

Material markup _____

Sales Tax _____

Total estimated cost _____

ESTIMATING PROCEDURE FOR INSTALLATION, DESIGN/BUILD PROJECTS

Design/build projects require a thorough analysis of the site conditions, soil structure, drainage issues, utilities, construction issues associated with subcontractors, site logistics, and verification of takeoffs and designated landscape construction areas.

Estimating production hours is often the most challenging part of an estimator's job for design/build projects. Although the company's production rates can be easily accessed through its estimating database, allowances need to be made for unforeseen circumstances associated with new construction sites if the contract does not have a **contingency clause** that stipulates compensation for complications such as severe weather or construction site delays. It is also imperative that the estimator thoroughly review the job specifications and architectural plans to ensure that compliance will not cause unusual financial risks, such as there would be in the installation of a roof garden with inadequate accommodations for drainage and weight.

Review of job specifications

Job specifications include such information as the size of materials (tree caliper, crown diameter) and installation standards (diameter of planting hole, elevation of root ball above grade, staking system, backfill mix). The specifications accommodate the landscape plans and are provided by the landscape architects based on industry standards, such as those provided by a trade association (e.g., Landscape Contractors Association MD, DC, VA), as well as those that are associated with municipalities. An example of exterior plant installation specifications for tree installation is shown.

Estimating

Tree installation (balled and burlapped)

A Planting pits should be dug with vertical sides in friable soils or angled outward and scarified in heavy soils.

B The depth of the pit should allow for 1/8 of the root ball to be above grade.

C There should be a minimum of 9" between the pit side walls and the root ball.

D Percolation tests are required in situations where poor drainage is anticipated. The test involves filling a 12" × 18" hole with water and assessing the amount of water remaining after eight hours. If excess water remains, a drainage system is required.

E Backfill mix shall consist of 3/4 native soil, 1/4 organic matter, and organic fertilizer. Soil tests are required to determine if any other additives are necessary.

F Remove the burlap from the upper 50% of the root ball and roll it back to the edges of the root ball. Remove all rope or synthetic twine from the entire root ball.

G Thoroughly mix the backfill and fill 50% of the tree pit, lightly tamping the mix during the process. Fill the remainder of the pit with the balance of the mix.

H Leave the top of the root ball exposed.

I Use the mix to form a saucer around the root ball.

J Mulch the top of the root ball, not to exceed a 3" depth. Taper the mulch away from the base of the trunk.

K Fill the saucer with water and follow with more water if the conditions are dry.

L Stake and/or attach guy wires for support if necessary.

Compilation

The compilation of an estimate involves the determination of:

I **Direct costs:**

    Material at wholesale cost

    Labor as actual hours, **crew average wage (CAW)**, or hourly production wage rate

Foreman	$21.91
Leadman	$15.25
Laborer	$12.95
Laborer	$12.95
	$44.06
CAW = $44.06 / 4 = $11.02	

Estimating

Labor burden as a percent of the hourly wage:
  Federal income tax
  Social security tax
  Medicare
  Federal unemployment tax
  State income tax
  State unemployment tax
  Workers' compensation insurance
Equipment costs based on hours of use and assessed cost per hour
Subcontractors at actual cost

II **Indirect costs**, encompassing those items necessary for production completion but not directly involved in the actual production phase. Silt fences would be an example in a construction project. Labor associated with loading and unloading materials and equipment is another example. Other examples of indirect costs include uniforms, permits, soil tests, temporary fencing, dump fees, travel time, supervisor hours (if not accounted for in direct labor), and watering plants in the staging (holding) area.

III **Markups**, percentage increases applied to specific costs for recovery of overhead costs. They are applied to:

   Labor to recover costs associated with supporting labor:
     Vacation compensation
     Sick leave compensation
     Holiday compensation
     Liability insurance
     Medical insurance
     Pension plan
     Training
     Uniforms
     Small tools
   Materials to recover overhead costs associated with handling and maintenance, loading and unloading

IV **Profit**, a percentage of the total job cost (see Chapter 3, "Profitable pricing")

   Profit = Breakeven cost* × complement factor **
   * Direct costs + overhead markup
   ** Complement factor is derived from Table 3.3, Profit Markup Factors

Estimating

Example

> Breakeven job cost = $12,000
> Profit (10%) = $12,000 × .1111 = $1,333.20 (complement factor of 10% = 10/90, or .1111)
> Selling price = $12,000 + $1,333.20 = $13,333.20
> Profit percent = $1,333.20/13,333.20 = 10%

An alternate method of finding the final selling price involves defining the breakeven price as a portion of the total selling price. Since the breakeven price plus the profit margin equals the final selling price, the breakeven price represents the portion of the final selling price that is equivalent to 100% less the profit percentage (called the inverse factor – see Chapter 3). Divide the breakeven price by the inverse factor (expressed in decimals), to derive the final selling price.

Typical net profits in the landscape industry are 6–10% for commercial maintenance jobs, 3–5% for commercial bid construction, and 10–20% for residential design/build projects. These percentages are contingent upon several factors:

1. Job size – smaller profit percentage with larger commercial jobs
2. Production efficiency – job cost management
3. Negotiated vs. bid contracts – generally larger profit percentages with negotiated contracts
4. Market competition
5. Weather/construction delays – resulting in labor production overruns
6. Bottom line – what the company needs for profit, growth, employment compensation, retiring debt

Travel time is also an important factor in estimating the job, since it allows managers to schedule the jobs within a realistic time frame. Scheduling allocates work hours and equipment on a daily basis from job initiation to completion and applies to the transitional phases of an extensive installation project as well as services provided in a maintenance contract.

The landscape construction format (Table 4.4) illustrates the components of a plant installation **assembly**, which represents a compilation of the operations associated with a specific phase of the landscape construction contract.

Another example of landscape construction assembly is **site preparation**. This process includes soil amelioration, which incorporates soil amendments into the native soil, grading, and demolition, which may include removal of plant materials or hardscape features such as walkways, driveways, or old patios.

Estimating

Table 4.4 Landscape Construction Estimate Format

Project Number	Assembly	Takeoff Quantity	Production Rate	Labor Hours	Labor Cost/Hr	Labor Cost
123	Planting trees *Platanus* × *acerifolia* "Bloodgood"	54 – 5"	6 hr/ea with equipment	324 hr	$13.79	$4,467.96
	Pre-emergent	3.91 lb				
	Fertilizing	35.10 lbs				
	Mulch rings	2.36 cyd	1.18 cyd/hr	2.00 hr	$13.79	$27.60
	Tree stakes (36")	162	6.67 ea/hr	24.30	$13.79	$335.10
	Hose	162 ft				
	Wire	1,620 ft				
	Backfill	8.10 cyd				

Project Number	Assembly	Material Cost/Unit	Material Cost	Total Cost
123	Planting trees *Platanus* × *acerifolia* "Bloodgood"	540.00	29,160	$33,627.96
	Pre-emergent	.77/lb	3.01	3.00
	Fertilizing	.07/lb	2.46	3.00
	Mulch rings	12.50 cyd	29.50	57.00
	Tree stakes (36")	.58 ea	93.96	429.00
	Hose	.30/ft	48.60	49.00
	Wire	.04/ft	64.80	65.00
	Backfill	8.50/yd3	68.85	69.00

Estimate totals:

Labor _____

Materials _____

Travel time _____

Supervisor cost _____

Material tax _____

Extra watering time _____

Freight – plants _____

Labor overhead markup _____

Material overhead markup _____

Subcontractors _____

Subtotal _____

Profit _____

Total estimate _____

Estimating

COMPUTER SYSTEM INTERFACE

Many computerized estimating programs are available to the landscape industry. An estimating system increases the efficiency of generating estimates and the accuracy of those estimates. Data entry includes the job site measurements, quantities, and services/operations. The estimating program generates the costs and production rates from its database and adds in the markups and profit. The database is updated on a regular basis to account for increases in labor costs, labor burden, and material costs. The benefits of a computerized estimating system are enumerated below:

Benefits

Enhances consistency and accuracy
Increases bid volume
Provides data for job costing
Gives flexibility in accounting interfacing
Provides database for generating project status reports

Essential elements

Supports staff with accurate and prompt data entry
Provides most current cost figures:

- Wages – crew average wage
- Labor burden
- Production standards
- Unit material costs
- Equipment hourly rates
- Overhead costs

TIME AND MATERIALS ESTIMATES

It is fairly common for landscape contractors to engage in work that was not included in the original contract, such as one-time leaf removal, the installation of a paved walkway, or the addition of a water feature to a design/build project. In these instances the contractors would bid the additional work as a **time and materials estimate** based on the labor and material

Estimating

costs required to complete the landscape operation. The pricing process is the same as in other estimates, but a portion of overhead is billed as a direct cost with labor because this lowers the markup for recovery of overhead. A lower markup enables a contractor to be in a more competitive position to secure the additional work.

The client and contractor agree on the pricing structure prior to the initiation of the operation. All costs associated with the additional work are listed and direct billed (invoiced). The hourly labor rate is inclusive of labor burden and overhead such as vacation and holiday compensation, sick leave, training, and so on, all of which are calculated as a percentage of the base hourly rate. A $12/hr supervisor with overhead and labor burden may therefore be billed at $18.34 (Table 4.5). A time and materials rate sheet is shown in Table 4.6. The labor rate and materials are specified for the landscape operation.

Table 4.5 Time and Materials Hourly Rate Components*

T&S TODD & SON LANDSCAPING

Labor Cost Components	Supervisor	Laborer
Base hourly rate	$18.00	$10.50
Vacation compensation	1.14	.63
Holiday compensation	.46	.32
Sick leave	.23	.14
Social Security tax	1.52	.84
Federal unemployment tax	.02	.01
State unemployment tax	.28	.17
Workers' compensation tax	1.34	.74
General liability insurance	1.06	.63
Medical insurance	1.30	.74
Pension plan	.45	.00
Employee training	1.27	2.10
Uniform	1.13	1.37
Small tools	.99	1.26
Total direct labor costs	$29.19	$19.45

* Components and their percentage of hourly rate are contingent upon company policies and state tax rates. Labor overhead and pre-tax profit are added to the labor hourly rate.

Source: *Pricing for the Green Industry*, 2012, 3rd ed., Frank Ross, Ross/Payne & Associates Inc., National Association of Landscape Professionals, Herndon, VA

Estimating

Table 4.6 Time and Materials Rate Sheet

T&S TODD & SON LANDSCAPING

Billing Information			Job Information		
Bill To:	RaeCo General Contractors		Date of Job Completion:	May 24, 20017	
Address:	3644 Boston Street		Job Site:	Lamco Corporate Park	
College Park, MD 20742	Job #:		43215		
Phone	(301) 555-7834		Contact Person:	Beth Paige	

Estimated Hours	Actual Hours	Quantity	Description	Unit Cost/Hour	Total Cost
2.0			Front-end loader	$65	
5.0			Labor	$30	
2.0			Supervisor	$45	
4.0			Equipment operator	$35	
		10 yd^3	Hardwood mulch	$25/yd^3	

SUMMARY

What is an estimating system?

An estimating system generates a price for all of the costs associated with the production of landscape operations. It uses the company's financial database to determine hourly production rates, overhead markups, and profit. The total estimated price is the bid that is submitted to the client. The estimate also serves as a job cost budget.

How are maintenance jobs estimated?

Maintenance jobs are estimated on the basis of the frequency of specified landscape services. Job site area measurements determine the labor hours required to perform the services. Company production rates provide the basis for estimating the hourly labor costs for the maintenance services. Material costs are estimated based on the quantity used. Overhead markups and profit are derived from budget projections.

Estimating

How are installation jobs estimated?

Installation jobs are estimated on quantity and area measurements derived from landscape plans and/or job sites. Assemblies compile the components of the installation production. Production rates are derived from the estimating database along with hourly cost rates, overhead markups, and profit. Large commercial installation jobs incorporate contingency fees for unforeseen circumstances and a risk factor percentage for unexpected site conditions.

What is the distinction between direct costs and indirect costs?

Direct costs are all costs associated with job production, including labor, labor burden, materials, and equipment. Indirect costs are related to jobs but are not production oriented, such as job permits, travel time, warranty expenses, small tools, and uniform expense.

What is the relationship between the estimating system and job cost management?

Estimates project all costs associated with landscape operations, including the number of hours of labor required to produce all aspects of the operations. An estimate therefore becomes the budget for managing landscape jobs. Managers are able to incorporate the projected costs into daily and weekly management reports for field supervisors.

KNOWLEDGE APPLICATION

1. What can an estimator do to ensure that the final estimate is an accurate assessment of the production costs?
2. How would you establish production standards for mowing residential properties?
3. Using the production rate tables in Tables 4.1 and 4.2, calculate the total production hours for:

    a Power stick edging, 3,000 linear feet, 24 weeks (1×/week)
    b Grading 5,000 SF with a tractor
    c Planting 10 flats of 2-1/4" ground cover

4. Differentiate between negotiated and non-negotiated contracts in terms of job estimates.

5

Financial management

CHAPTER OBJECTIVES

To gain an understanding of:

1. The components of balance sheets and income statements
   a. Balance sheets
      i. Current assets
      ii. Fixed assets
      iii. Accounts receivable
      iv. Accounts payable
      v. Current liabilities
      vi. Long-term liabilities
      vii. Accrued taxes and salaries
      viii. Owner's equity
   b. Income statements
      i. Direct costs
      ii. Gross profit margin
      iii. Indirect costs
      iv. General and administrative overhead
      v. Net pre-tax income
2. The financial implications of balance sheets and income statements
   a. Balance sheets reflect financial status
   b. Income statements reflect profitability
3. Job cost management systems
   a. Tracking direct production costs

Financial management

    b Monitoring actual versus projected budget
    c Generating updated job status reports
4 Job cost management
    a Labor production hours
    b Material invoiced costs
    c Projected completion costs (labor, subcontractor and material costs)
5 Management of profit centers (divisions/departments)
    a Increase production efficiency
    b Reduce overtime
    c Establish production standards
6 Contract management responsibilities
    a Site management
    b Mobilization
    c Monitoring labor production costs
    d Monitoring material purchases
    e Subcontractor management
    f Meeting specifications
    g Scheduling client meetings
    h Monitoring accounts payable and receivable
    i Change orders
    j Submission of percentage completion invoices
    k Warranty management
    l Demobilization

KEY TERMS

balance sheet	job cost management	solvency
cash flow report	owner's equity	work in progress schedule (WIP)
financial management	purchase order	
income statement	retainage fee	

A financial reporting system interfaces with all business operations and enables management to monitor the pulse of the company. A financial library is established that categorizes every financial transaction that occurs in a company, which is used as a reference for the development of budgets, pricing structures, and estimating systems. These are the building blocks for financial management of projected revenue and expenses, pricing structures that encapsulate direct and overhead costs, and estimating systems that generate profitable sales.

    Compilation of a company's financial information into two key reports, the *balance sheet* and the *income statement*, provides management with heart monitors. Equally if not more

Financial management

important than monitoring the heart is monitoring the pulse. This is accomplished with a *cash flow report* that determines whether the blood is flowing upstream or downstream and at what rate. All of these financial reports are important tools that management uses to ensure a company functions in a profitable mode. The financial information generated in these reports enables management to identify costs that are exceeding budget projections and to make informed decisions that will enable the company to meet its financial obligations.

BALANCE SHEET

The **balance sheet** is the most important document produced by an accounting system. It indicates the financial status of a business on a specified date and establishes the company's **solvency**, or its ability to meet its financial obligations. As the name implies, the balance sheet reflects a balance between assets and liabilities.

A balance sheet application can be illustrated with a personal finance situation. A couple decides that they want to buy a bigger house and purchase an SUV. First they have to add up their assets: the cash in their bank and checking accounts, stocks, bonds, home equity (difference between mortgage balance and market value), furniture, and any other property that has monetary value. Then they subtract the total of all their liabilities – credit card debt, outstanding bills, mortgage balance, and other loan balances – from their total assets. The resulting dollar amount represents the current net worth of the couple, which is referred to in business terms as **owner's equity**. The couple's equity will indicate how much liability they can assume with their desired purchases

Table 5.1 Balance Sheet Components

Assets	Liabilities
Assets represent everything of economic value, which includes current and fixed assets:	Liabilities are financial obligations:
Current assets – include cash in bank accounts, short-term investments, money market accounts, and any other investments or assets that can be converted to cash quickly, such as inventory and securities. Inventory items, whether plant materials, chemicals, or hardscape materials, represent items that haven't been sold. Their value is assessed at wholesale cost.	**Current liabilities** – are those that are due within a year or less, e.g. vendor invoices, taxes.
	Accounts payable – are monies owed to vendors, utilities, and other expense sources associated with operational expenses.
Accounts receivable – represent revenues earned and billed but not collected. They remain categorized as current assets until they exceed a certain number of days outstanding, generally more than ninety days. Those accounts that remain outstanding are listed as bad debt.	**Accrued taxes and salaries** – are those that have not been paid as of the date the balance sheet is generated.
	Long-term liabilities – are those that are due beyond a year, e.g., mortgages, vehicle and equipment loans.
Fixed assets – are of a more permanent nature: buildings, equipment, and real estate, all of which have value but are not intended for liquidation. Equipment assets all have an annual depreciation (tax allowance) that is deducted from their value.	**Owner's equity** represents the amount of money that remains after liabilities are subtracted from assets.

of a new home and an SUV. These individuals – as well as businesses – also have to take into consideration their current cash flow to meet their monthly financial obligations (liabilities).

In principle, this p xsed in preparing a company's balance sheet (Table 5.2).

Table 5.2   Balance Sheet Format

BradCo Landscaping Company
March 31, 2017

Assets

Current assets
Cash & securities	$30,000	
Accounts receivable	$150,000	
Inventory	$60,000	
Total current assets		$240,000

Fixed assets
Land	$75,000	
Building and improvements $250,000		
Less: Accumulated depreciation −$95,000	$155,000	
Furniture and fixtures $30,000		
Less: Accumulated depreciation −$16,000	$14,000	
Vehicles and equipment $375,000		
Less: Accumulated depreciation −$165,000	$210,000	
Total fixed assets		$454,000
Total assets		**$694,000**

Liabilities and owner's equity

Current liabilities
Accounts payable	$50,000	
Accrued taxes	$9,000	
Accrued salaries	$175,000	
Notes payable (due December 2016)	$50,000	
Total current liabilities		$284,000

Long-term liabilities
Notes payable (due June 2016)	$105,000	
Total long-term liabilities		$105,000
Total liabilities		**$389,000**

Owner's equity
Common Stock (dividends)	$100,000	
Retained earnings	$205,000	
Total owner's equity		**$305,000**
Total liabilities plus owner's equity		**$694,000**

Financial management

The balance sheet is based on the fundamental accounting equation:

Assets = Liabilities + Owner's Equity

The assets are balanced by the liabilities and owner's equity. Liabilities are financial obligations that are paid from current assets, and the difference between the liabilities and assets represents owner's equity. BradCo Landscape Company's owner's equity, $305,000, is the difference between the company's assets and its liabilities ($694,000 − $389,000). Total assets ($694,000) equals total liabilities plus owner's equity ($694,000).

Balance sheets are generated on a quarterly basis as indicators of the financial health of the company based on the ratio of liabilities to assets. The end-of-year balance sheet generated on December 31 reflects whether the company was fiscally responsible in managing its short- and long-term liabilities as indicated by the year-end ratio of assets to liabilities.

Today, banks are much more conservative in issuing loans and will require a ratio of 2:1 ($2 of assets for every $1 of liability) as an indication of a company's financial stability and as a qualifying factor for a credit line or loan. In this example, the BradCo Landscape Company is in good financial health for the first quarter of 2016. Its assets ($694,000) are greater than its liabilities ($389,000), so if the business closed its doors tomorrow, it would be able to liquidate its assets and meet all of its financial liabilities.

In further analysis, however, management would note that its total current assets ($240,000) are insufficient to meet its total current liabilities ($284,000). The concern is that if the company's sales slumped due to weather or contract cancellations, there would be insufficient funds to meet the company's current liability obligations. Management would therefore consider building up a larger cash reserve and ensuring a line of credit in the event of negative cash flow (expenses exceeding revenue). The collections department would also be notified to keep close tabs on accounts receivable. Although accounts receivable are assets, unpaid accounts become delinquent after ninety days and will be accounted for as bad debt.

To assess financial performance, balance sheets are always compared to past years to determine consistency or inconsistency in the management of assets and liabilities. The balance sheet also reflects the financial impact of a new division/profit center, such as sports turf maintenance, irrigation maintenance, or tree service, because of the liabilities that would be associated with its startup expenses.

Here are some benchmarks (standards) that are recommended for balance sheets:

Debt to assets	Long-term debt dollars/ Net fixed asset dollars	40% or less
Asset turnover	Revenue dollars/ Total asset dollars	375% or greater
Current ratio	Current assets (Cash + Accounts Receivable)/ Current liabilities	200% or greater

Financial management

INCOME STATEMENT

In contrast to the balance sheet that reflects a company's overall financial status, an **income statement** reflects the profit that is earned from revenue during the designated financial period. The term *income statement* is used interchangeably with *profit and loss statement*. When managers refer to the "bottom line" they are referring to the amount of profit retained from income revenues after costs and expenses have been paid.

The balance sheet reflects a company's financial status within a specific period. The income statement reflects the profitability of a company's business operations within a specific period of time. A balance sheet is a snapshot while an income statement is a motion picture.

In order to achieve profit goals, management must be proactive and monitor weekly and monthly income statements to detect deviations from the budget. By analyzing income statements, managers can determine what adjustments need to be made to attain their respective budgeted targets.

The annual income statement as shown in Table 5.3 provides an overall summary of a company's financial performance, which can be evaluated on the basis of benchmarks achieved in previous years. The percentages, similar to production time standards, provide a means of measuring costs against a baseline.

Table 5.3 Annual Income Statement

Columbia Landscape Maintenance Company Annual Income Statement, 2016 Sales Volume $580,000		
	Percentage of Sales	$
Net sales	100%	$580,000
Direct job costs		
Direct labor	30.0	174,000
Direct labor burden	4.0	23,200
Material costs	16.0	92,800
Subcontractors	5.1	29,580
Other direct costs	1.4	8,120
Total direct job costs	56.5	($327,700)
Gross profit margin	43.5	$252,300
Indirect costs		
Indirect labor	.5	2,900
Equipment rental	1.2	6,960
Fuel & oil	2.7	15,660
Equipment/vehicle insurance	1.0	5,800

(Continued)

Financial management

Table 5.3 (Continued)

Columbia Landscape Maintenance Company
Annual Income Statement, 2016
Sales Volume $580,000

	Percentage of Sales	$
Equipment/vehicle maintenance	3.6	20,880
Tools & supplies	.9	5,220
Miscellaneous	.5	2,900
Total indirect overhead	**10.4**	**($60,320)**
General & administrative overhead		
Marketing	0.6	3,480
Depreciation	3.8	22,040
Insurance – medical & life	1.1	6,380
Insurance – liability	.8	4,640
Insurance – workers' compensation	1.0	5,800
Office expense	1.3	7,540
Payroll taxes	1.2	6,960
Profit sharing/pension	.4	2,320
Rent	1.7	9,960
Salaries – officers/owners	8.0	46,400
Salaries – administrative	4.0	23,200
Salaries – sales	.2	1,160
Communication system	1.3	7,540
Travel & entertainment	.4	2,320
Utilities	.4	2,320
Miscellaneous	2.1	12,180
Total general and administrative overhead	**28.3**	**($164,240)**
Net income (profit) before taxes	**4.8**	**$27,740**
Pretax profit + Owner's salary ($46,500)	12.8	$74,240

Direct costs, which are associated with production, labor, and materials, are the primary focus of the profit center (divisions) income statements. The efficiency factor reflects the amount of time expended to complete production operations. Landscape companies continuously strive for production efficiency to reduce labor costs, the major constituent of direct costs. The effect of these efforts will be reflected on the income statement. Non-billable labor is one cost that can be reduced easily, since it is often associated with downtime or nonproductive time. If the labor percentage exceeds the budget, it may be indicative of unproductive labor costs, which may be associated with delayed crew departures, excessive travel time, or fueling and loading inefficiency. Other factors such as weather and underestimates may also be related to labor budget variances.

Financial management

Fixed overhead costs, such as rent, utilities, administrative salaries, advertising expenses, and equipment maintenance, are incurred regardless of the sales volume and must be astutely managed to ensure they do not have a negative impact on the company's profit.

Software accounting systems enable real-time income statements to be generated daily, weekly, and monthly to monitor the bottom-line status of the entire company, individual branches, and each profit center. Regional offices of large corporations utilize the Internet to compare their financial performance against one another.

Think of the income statement as a report card. The company's grade is reflected on the bottom line. A spreadsheet format as shown in Table 5.4 is set up to provide a comparison of current company performance against the budget for the period and year to date. The data components of each cost category would be entered and compared to budgeted figures.

Table 5.4 Income Spreadsheet Format

Current Period:						
	Actual		Budget		Variance	
Net Sales	$	%	$	%		
Direct Job Costs						
Total Direct Costs						
Gross Margin						
Indirect Overhead						
Total Direct Overhead						
General & Admin. Overhead						
Total General & Admin. Overhead						
Net Profit						
Year to Date:						
	Actual		Budget		Variance	
Net Sales	$	%	$	%		
Direct Job Costs						
Total Direct Costs						
Gross Margin						
Indirect Overhead						
Total Direct Overhead						
General & Admin. Overhead						
Total General & Admin. Overhead						
Net Profit						

Financial management

CASH FLOW REPORTS

A **cash flow report** indicates the source and current amount of cash received, the current cash disbursements, and the cash balance. Regardless of the sales volume and profit of a company, unless it operates in a positive cash mode, it will be in financial trouble. Companies generate profit and loss statements (income statements), but these don't tell the whole story. On paper, these statements may show that the company is profitable on generated sales, but they don't say how much cash is available to pay bills, salaries, loan payments, and so forth. Profit is just a number on a page until the cash is in the bank.

If cash flow is reduced because of delinquent accounts receivable, late invoicing, or loss of business, a company can find itself financially overextended. In other words, not enough cash is coming in to pay the bills. In other situations, design/build and landscape construction companies can easily go into negative cash flow when working on large projects involving high labor and material costs, which are paid in advance of final billing. Commercial construction companies generally receive partial payments based upon the percentage of completion of various phases of the job. However, regardless of the partial payments, costs can still exceed cash flow from these projects at times over the course of the project. A company has to be careful not to overcommit to such jobs unless it uses a line of credit or has sufficient cash reserves to finance the upfront costs.

Maintenance contracts are prorated over a twelve-month period. Therefore, companies receive a steady monthly cash flow regardless of the season. In the late fall and winter months, production costs are low due to seasonal labor layoffs and reduced labor associated with the services required at that time of year. Therefore, the company is in positive cash flow since the income revenues exceed the cash outflows. However, when the spring season begins, material and labor costs rise while the revenues remain constant, thereby causing a negative cash flow. A maintenance company whose fiscal year begins in April will experience negative cash flow for approximately six months (in regions with forty-week seasons). A weekly cash flow report will raise a red flag when it appears that cash flow is not meeting projected budget figures. Cash reserves or a short-term loan will enable a company to meet its financial liabilities. During periods of positive cash flow, any excess cash remaining after paying expenses can be invested in short-term securities.

Cash flow reports (Table 5.5) provide a means of tracking actual income and direct expenses against monthly projections. Since labor is the primary expense of maintenance companies, these reports emphasize projected payroll for the period against projected income. This financial tool is the key that keeps the doors open and the trucks rolling.

Financial management

Table 5.5 Cash Flow Report

	January	February	March
Current balance			
Cash received			
Sales			
Accounts receivable			
Other			
Total cash received			
Cash disbursements			
Fuel			
Insurance			
Interest			
Office expenses			
Inventory – fertilizer			
Inventory – plants			
Payroll			
Payroll taxes			
Professional fees – legal			
Lease – trucks			
Rent			
Utilities			
Other			
Total cash disbursements			
Cash balance			

Financial management is proactive engagement with financial reports, which involves analyzing information on the reports and implementing adjustments wherever necessary to meet budgeted projections.

- **Balance sheets** provide current information on the company's financial status
- **Income statements** reflect profitability status
- **Cash flow reports** indicate whether the cash on hand is enough to pay the bills

These three financial entities provide a quantitative analysis of the company's current financial status. Monitoring of revenues, costs and expenses, and cash flow keeps managers informed and prepared to meet financial challenges and maintain profitable company operations.

Financial management

JOB COST MANAGEMENT

Profit centers are where the company's financial outcome is determined. An accurate statement in this regard is "Front line is bottom line." **Job cost management** entails monitoring direct costs – labor and materials – against job budgets. Managers are constantly assessing costs against projected revenue. Each day on the job is equated to work hours, with the actual expended time being compared to the budgeted amount of hours (derived from the job estimate). Field reports (Table 5.6) are generated to provide supervisors with projected hours for the scheduled service or work phase. Weekly summaries indicate which job operations are within budgeted hours and which ones are exceeding projections. In instances where labor is over budget, an analysis of the job reports indicates where the overruns are occurring.

The job cost management system relies on software programs to produce reports that enable managers to track production costs from start to finish. Knowing the current job status allows for adjustments to be made on a job to put it back on a profitable track.

The information derived from job cost management systems enables companies to bid future work with more accuracy. Labor costs for hardscape installation can be broken out from plant installation to determine markups based on productivity levels – in other words, to determine the hourly rates or standards based on production times for specific types of installations. Problems with excessive production time that arise during jobs can be more readily resolved in the future by recognizing a more efficient means of working to increase profitability. Examples of efficient working methods would be having a chipper available for large pruning jobs, using leaf vacuums for leaf removal, or utilizing a compact utility loader for more extensive residential installations.

Table 5.6 Job Field Report

	Job # _____		
	Location _____		
	Start Date _____		
	Completion Date _____		
Materials	Quantities	Equipment	Hours
Operations/services: Hours			
Contact			
Contact information			
Name _____			
Telephone _____			

Financial management

The estimate

The first layer of a job cost system is the estimate, which projects production hours and costs, material quantities and costs, equipment costs, subcontractors' costs, and other direct costs associated with the job. The costs and quantities are broken out by the service or phase of the job. Once the estimate is accepted, it becomes the budget for the project/job. Each phase/service of the job has cost breakdowns that assist project/account managers in scheduling crews and managing their production hours. Most importantly, keeping track of the budgeted hours lets field supervisors know how much time they have to get the job done. Labor budgets (Table 5.7) are set up on a week-to-week basis, and management tracks actual hours to budgeted hours. Adjustments are made due to change orders (additional work beyond the contract), weather factors, or any other factors that would cause deviations from the production schedule.

Job cost accountability is achieved through input of daily job reports (Table 5.8). In Chapter 8 we will see how this task is facilitated by technology such as GPS (global positioning satellite) systems, phone applications, and integrated software programs. The computerized field reports are uploaded to the company server and directed to payroll and accounting with the job codes and respective production hours. Material entries are also recorded and checked with invoices to ensure that the budget quantities are being directed to the job.

Table 5.7 Weekly Labor Production Report

Job/Client Name			Week Ending	
Operation/Service	Budgeted Hours	Actual Hours	% Completion	M T W Th F
Total				

Financial management

Table 5.8 Daily Job Report

| Job Name _____ Date _____ |
| Supervisor _____ |
| _____ |
| Employee name _____ |
| Hours |
| M: |
| T: |
| W: |
| Th: |
| F: |
| Total hours _____ |
| Operation Code _____ |
| Job Account _____ |

Table 5.9 Inventory Stock List

Inventory Stock list				
Item Description	Qty. Available	Qty. Min.	Qty. Max.	Qty. on Order

Inventory stock lists provide the current status of materials on hand and the quantities that need to be maintained.

Source: *Blueprint for Success*, National Association of Landscape Professionals (formerly Professional Landcare Network), Herndon, VA, in cooperation with Caterpillar Inc.

Inventory tracking

Another important cost factor is associated with inventory (Table 5.9). An inventory system tracks where the materials go. An inventory manager controls inventory outflow and maintains stock of required material quantities.

Project managers are required to sign for all materials taken from inventory, whether plants, fertilizer, or fieldstone, and assign them job numbers. Too often, inventory items are used on jobs without any record of job allocation. Negligent recording of inventory items prevents recovery of purchase costs and profit and distorts job cost reports with inaccurate material costs.

Financial management

Material costs

Material costs are kept current by posting invoices to jobs as soon as they are received. Receipt of shipments (shipping receipts) is critical to the accuracy of these postings to ensure that the quantities and type of materials correspond to the **purchase orders** that stipulate the quantity of items ordered, the vendor, and the authorizing manager (Table 5.10).

Job cost reports provide information on actual costs versus budgeted costs:

1. Production hours – current and projected at completion
2. Material costs
3. Projected total cost at completion
4. Total costs in each category

At any given time, a project manager can access the total costs of any one job and compare it to its budget projections. In the example given in Table 5.11, labor hours are slightly under budget for the month of April and to date for the UMD installation project. Sammy, the project manager, notices that the plant costs are 9% over budget. He checks with the plant purchaser to determine the discrepancy from the budget projection. After checking the

Table 5.10 Purchase Order

Purchase order

Date _____ Job Name _____
Vendor _____ Job Number _____
Purchase Order Number _____ Account Number _____
Authorization _____

Date Ordered	Vendor Phone	Vendor Contact	Date Requested	Ship Via		FOB

Quantity	Unit Size	Description	Unit Price	Extension	Invoice	Invoice Date

Source: *Blueprint for Success*, National Association of Landscape Professionals, Herndon, VA, in cooperation with Caterpillar Inc.

Financial management

Table 5.11 Job Cost Report

BradCo Landscaping Company
Job Cost Report
UMD Plant Science Building
April 20016

	Actual	Budget	% Budget	YTD Actual	TTD Budget	% Budget
Direct costs						
Labor hours						
Planting	155	162	96%	287	295	97%
Staking	45	45	100%	63	63	100%
Mulching	10	8	125%	22	16	137%
Total hours	**210**	**215**	**98%**	**372**	**374**	**99%**
Material Costs						
Plants	$58,000	$53,000	109%	$73,000	$68,000	109%
Tree stakes	$840	$840	100%	$1,003	$1,003	100%
Mulch	$206	$206	100%	$335	$335	100%
Backfill	$310	$310	100%	$460	$560	100%
Total costs	**$59,356**	**$54,350**	**109%**	**$74,798**	**$69,798**	**107%**

plant invoices, they determine that the cost discrepancy is due to a higher unit cost for some "Bloodgood" London plane trees. This is not uncommon if there is a limited supply of a plant variety or if the item is purchased in small quantities and thus is not eligible for a quantity discount.

Production hours

In addition to computerized job reports, maintenance companies prefer to have job boards (Table 5.12) posted that record the weekly status of production hours, actual versus budgeted, for each scheduled service and client property. At a glance, field supervisors can see how much time they have remaining for each job. Adjustments can be made in routing/scheduling jobs to redirect labor hours for completing services that have been delayed, such as mulching when early spring causes mowing to begin earlier. Management might subcontract out to mulching companies that use mulching trucks to complete that phase of the contract.

Financial management

Table 5.12 Monthly Maintenance Service Tracking

Job Site	\multicolumn{5}{c}{February 2017 Weekly Hours}			
	1	2	3	4	Total Actual	Total Budget	% Budget
Tupper Condos Mulching Pruning							

Table 5.13 Schedule Board

Jobs on Schedule						
Jobs to be Scheduled	Monday	Tuesday	Wednesday	Thursday	Friday	to be Billed

Job cost variances commonly occur with both landscape maintenance and construction jobs due to weather, site conditions, delayed material deliveries, and so on. On the other hand, the job may be progressing ahead of schedule due to favorable weather conditions and production efficiency. In both cases, the budget versus actual production costs will reflect the variances associated with the respective job status, expressed as a percentage of the budget cost projections. As previously mentioned, each of these reports becomes a reference document for estimating similar jobs in the future. Collectively, the reports comprise the profit center's income statement that will reflect its profitability.

Landscape construction companies use scheduling boards (Table 5.13) to post start dates and completion dates. The schedule board may have three to four months of postings.

Financial management

Table 5.14 Schedule Board Job Format

Job No. _____ Job Client _____ Start Date _____ Completion Date _____
Contact _____ Telephone No. _____ Job Location _____
Directions _____
Budgeted Direct Job Costs $ _____ Budget Gross Profit _____
Materials   **Quantity**   **Unit Cost**   **Total Cost**

Depending on weather and other factors, the board will be updated with revised start and completion dates. Pertinent information for each job will be listed, such as directions, contact information, job number, location, estimated costs, gross profit, and budgeted direct job costs (materials, labor, subcontracts). This update resource provides an invaluable communication tool between sales and production personnel. This information may also be accessed via software scheduling programs.

Job cost reports

Quarterly and annual job cost reports are issued for each profit center. The costs are compared to the previous year to assess the profit center's profitability. If the budgeted profit is not being met, an analysis of costs will indicate where the overage occurs. This cost summary will also indicate whether the productivity level is sufficient and if production standards and pricing need to be adjusted for future jobs.

Tracking direct costs is what job cost management is about. The objective of job cost management is to monitor and manage. Keeping current with jobs in progress enables management to analyze current information and make adjustments that will increase bottom-line profit. In instances where costs exceed budgeted costs in spite of productivity efficiency, there may be a problem with the current estimating process. Businesses that succeed in today's competitive market do so by concentrating on improving productivity.

Establishing standards and providing resources such as equipment, training, GPS routing, and so forth leads to greater profitability. Managing operations within each profit center is the key to eliminating waste associated with "call-backs" and downtime. Labor waste is illustrated by Table 5.15, which demonstrates how much daily downtime costs the company in productive hours and dollars. The time lost may be associated with excessive travel time, fueling, late departure from the yard, breakdown in equipment, or return trips for job equipment or materials.

The loss in billable time is associated with production inefficiency. It's the job of management to determine where the inefficiencies occur and to take corrective actions to eliminate their recurrence. Regardless of how long a company has been in business, management

Table 5.15 Economics of Nonproductive Time

No. of Crews*	Billing Rate/Hour	15-Minute Loss in Billing Time $ Loss			
		Per Day	Per Week	Per Month	Annually**
1	$35	$8.75	$43.75	$175	$1,837.50
5	$35	$43.75	$218.75	$875	$9,187.50
10	$35	$87.40	$437.50	$1,750	$18,375.00
20	$35	$175.00	$875.00	$3,500	$36,750.00

* Three-person
** Based on 42 weeks of production
Per day = .25 × bill rate/hour
Per week = Per day × 5 days
Per month = Per week × 4 weeks
Annually = Per month × 10.5 months

never becomes complacent about job cost management. There is always room for improvement, and in the competitive environment of the landscape industry, it is essential that a company increase its productivity efficiency to enable it to deliver its services at a profitable and competitive price.

A job cost hour's summary sheet enables a quick analysis of labor hours throughout the term of a service contract. It can be cross-referenced to the scheduled services to assess production efficiency. In the report in Table 5.16, hours and dollars expended for direct labor costs are fairly close to budgeted figures. However, these differences still raise a red flag and warrant an analysis of the variances. Was it due to weather factors, equipment breakdowns, or errors in estimated budgets? Remember that direct costs refers to the labor hours that were applied directly to the job.

The indirect hours are 31% over budget, which equates to $48,178 ($335,692 − $287,514). The indirect hours include travel time for trips to the dump, pickup of materials, fueling time, time to load and unload the trucks, and the time it takes the crews to leave the yard. There is obviously waste or downtime, which needs to be identified and rectified.

Further exacerbating the labor hours is overtime. Overtime is necessary only when weather conditions, time constraints, or other extenuating circumstances (such as an insufficient labor force) prevail. The overrun in budgeted overtime is $116,400. A control valve for restricting overtime hour overflow is to require account managers to get approval from their branch manager prior to authorizing overtime expenditures.

Total labor cost overruns for the BradCo maintenance division through October are $319,185 ($154,607 direct labor cost, $48,178 indirect labor cost, and $116,400 overtime

Financial management

Table 5.16 Tracking Division Job Cost Labor Hours

BradCo Landscaping Company

BradCo Landscaping Company College Park Branch Maintenance Division
October 2017

	Week 1	Week 2	Week 3	Week 4	Total	Budget	% Budget	YTD Total	YTD Budget	YTD %
Direct hours	12,778	12,778	10,885	11,169	47,590	59,206	80.4	488,684	464,041	105.3
$	112,162	111,724	95,843	98,701	418,429	480,118	87.2	4,241,738	4,087,131	103.8
Indirect hours	311	152	197	205	864	760	113.6	29,752	22,700	131.1
$	3,085	1,961	2,457	3,064	11,288	8,709	129.6	335,692	287,514	116.8
Overtime	15,425	15,291	13,934	12,886	57,537	57,863	99.4	577,640	461,240	125.2

labor cost). The profit and loss statement for this division at this time would bleed red ink. Questions need to be asked and answers found to enable corrective actions to be taken.

The landscape industry generates revenues through the combined production of equipment and labor. Profit from the generated revenues will be maximized if the amount of production hours for both of the entities is minimized. This can be achieved with a job cost management system.

- Production is the process of performing a task.
- Estimating how long that task will take is also a process.
- Establishing production predictability is the key to recurring profitability.

CONSTRUCTION CONTRACT MANAGEMENT

Construction contract management utilizes job cost management to keep current with daily production costs. It is vital to both maintenance and installation contracts. It is particularly critical for project managers associated with large commercial installation projects. These projects are generally bid on a very competitive basis and therefore are managed on a tight budget with very little room for error. Production inefficiencies, schedule delays, or purchases that exceed budgeted costs all can adversely affect the profitability of the project. Project managers in these situations are responsible for:

- Site management – logistics, scheduling, safety
- Mobilization – equipment, staging holding areas for materials
- Personnel management – scheduling
- Work in progress scheduling

Financial management

- Monitoring labor production costs – tracking against budget
- Monitoring purchases – tracking invoices against budget
- Producing invoices – timely issuance
- Monitoring percent of completion – phases of construction
- Recovering retainage (commercial construction) – fulfilling specs on schedule
- Monitoring payables and receivables – tracking
- Warranty management – monitoring and scheduling
- Change order issuance (commercial construction) – additional work
- Inspection for quality control – meeting specifications

The project manager will maintain a **work in progress schedule (WIP)** for each project that summarizes the current production costs and what has been billed to date. The daily summary enables the manager to check job hours against budgeted (estimated) hours for every phase of the project. The WIP schedule also reflects the completion stage of the project. This information is vital for maintaining cash flow to meet payables to vendors and subcontractors. General contractors require monthly invoices from the landscape contractor that itemize what has been completed in the job phase – irrigation, tree installation, sod installation, paver installation, and so on. The landscape company is paid thirty days after the submittal of the invoice.

Contracts, particularly commercial contracts, have a clause specifying a **retainage fee** that is generally 10% of the total amount of the contract. This amount is retained by the client until the completed landscape installation is inspected and it has been verified that all of the specifications have been met.

Stringent contract management ensures:

- Profit
- Production efficiency
- Production coordination
- Communication with general contractor/client
- Goals will be set and met
- Priorities are set

The mutual goal of job cost management and construction management is to measure where the company is.

You are what you measure.[1]

Measurement is the key to proactive management and increased profitability.

Financial management

SUMMARY

Financial management involves monitoring the costs of delivering services and proactively implementing corrective actions to maintain profitability. Accounting systems provide real-time information for financial managers to generate reports and statements that summarize the current financial status of the company, its branches, and individual profit centers.

What are the components of balance sheets and income statements? What are the financial implications of balance sheets and income statements?

Balance sheet components include current and fixed assets, current and long-term liabilities, and owner's equity. The ratio of assets to liabilities indicates a company's financial stability and reflects its ability to meet its financial obligations.

Income statements, also referred to as profit and loss statements, consist of a report of current income revenues and the direct costs, indirect costs, and overhead expenses associated with those revenues. The income statement reflects profit or loss for a specific period based on the difference between revenues and expenses. Income statements indicate whether a company, its branches, and its profit centers are operating at a profit.

What are job cost management systems? What is job cost reporting?

Job cost management entails keeping track of day-to-day costs associated with production. Managers compare actual costs of labor and materials against budgets and determine where adjustments need to be made to attain budgeted goals. Job cost estimates serve as the individual job cost budgets.

Job cost reports provide the managers with current production information presented as a comparison between actual and budgeted labor production hours, labor costs, and material costs for the respective period, year to date, and as a percent of the budget.

What are the responsibilities of construction contract management?

Construction contract management involves maintaining a work in progress schedule for each project. Responsibilities include site management, mobilization, personnel management, monitoring direct costs against budget, generating invoices, monitoring percentage of completion, recovering retainage, monitoring cash flow, warranty management, issuing change orders, and quality control.

KNOWLEDGE APPLICATION

1. Contrast the differences between balance sheets and income statements in regard to:
   a. Their financial components
   b. Their financial significance
2. As a manager, how would you reduce direct costs associated with tree, turf, and seasonal color installations?
3. Based on the income statement information for the Terp Landscaping Company below, what is:
   a. The gross margin for each year?
   b. The net profit for each year?
   c. Your financial assessment of the company?

TERP LANDSCAPING COMPANY

	2015	2016	2017
Total sales	$1,520,000	$1,884,800	$2,544,500
Direct costs	819,000	1,005,000	1,323,000
Overhead	581,000	750,000	1,070,000

4. You have just received a weekly job cost report that indicates a $3,000 overrun in labor hours associated with a plant installation project in Washington, DC.
   a. What factors could possibly contribute to this cost overrun?
   b. What adjustments need to be made to get back on budget?

NOTE

1. Bruce Hunt, 2005, Former Vice President Brickman Group, Ltd. Personal Communication.

6

Project management

CHAPTER OBJECTIVES

To gain an understanding of:

1. Best management practices
   a. Maximize efficiency
   b. Standardize production procedures
2. Project management processes
   a. Scope assessment
   b. Planning
   c. Monitoring and controlling
   d. Execution
   e. Closure
3. Project management accountability
   a. Daily job reports
   b. Production efficiency
   c. Benchmarks
   d. Meeting production standards
   e. Actual versus projected budget

KEY TERMS

as-builts
best management practices (BMP)
closure
execution
lead time
minority/women's business enterprise (MBE/WBE)
monitoring and control
planning/pre-engineering
procedures
scope

Project management

Project management, as it relates to the landscape industry, is primarily associated with construction. There are certainly applications to maintenance/management contracts, but the procedures defined in project management are critical to the construction sector of the industry. Astute management is vital to the profitability of a construction company and is also a key to surviving economic downturns, particularly while managing large projects with lower profit margins.

Commercial landscape project managers are responsible for every aspect of the project, including supervision of subcontractors for all aspects of exterior installations – electrical, plumbing, concrete, storm water management, and so on. Landscape project managers need to be familiar with the associated construction standards and acquire a broad range of expertise in order to successfully manage all phases of a project.

Essentially, project management is the implementation of a set of **procedures**. A company compiles a database of successful procedures (inclusive of production standards) from previous projects, which are organized into standardized procedure modules that can be implemented in future projects of any size or monetary value.

Quality standards are a key component of a company's procedures. When these standards are followed, project managers can rise to the challenge of meeting deadlines and staying on budget. Delayed deliveries, incorrect plant installations, and site problems, such as poor drainage, soil compaction, a labor shortage, or a landscape architect's incorrect selection or rejection of plant materials, are just a few examples of problematic developments that can contribute to budget overruns.

The establishment of standardized project procedures becomes **best management practices**. Each step in the management procedures determines the success of the following steps. As we discuss these procedures, it will become evident how much information must be acquired by the project manager before any of the procedures are implemented. Examples are companies that have entered the green roof sector of the green industry, as well as those that install storm water management facilities such as biofiltration ponds and water harvesting cisterns. Each new project requires research and the development of new production procedures.

PROJECT MANAGEMENT PROCESSES

A structured sequence of management processes is followed before and after project implementation. It is a team approach that engages all parties associated with the project. The sequence is:

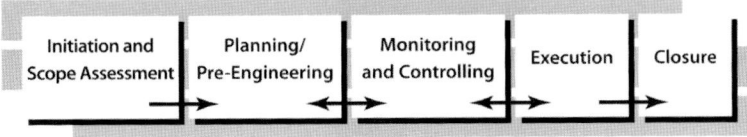

Project management resource: Project Management Institute, www.pmi.org
Certification, Project Management Body of Knowledge Manual

Project management

Initiation and scope assessment

Initially, the company management team assesses the **scope** of the project to determine what a project encompasses. Management then meets in what is referred to as a "turnover meeting" with the business developer (salesperson), branch manager, estimator, project manager, and superintendent, who discuss the important features of the project to ensure that everyone is on the same page.

The following aspects of the project are assessed before the project is implemented:

- Expertise – does the company have it in-house or does it need to be secured by either hiring experienced workers or outsourcing the work through subcontractors?
    Are there capable managers in-house and a dependable support team?
    Does the size of the project require a manager with more experience?
    Is there a management team that can work with the project manager to develop (new) procedures?
    An actual example of team support occurred with Outside Unlimited, a commercial landscape construction company in Maryland. The company bid on a five-acre green roof with a contract value of $3.5 million. Although they had no experience in green roof installations, they considered this as an opportunity to develop a new market niche. Their team worked with outside consultants to develop procedures for this unique project and met weekly to review and fine tune the processes. The end result was that Outside Unlimited became one of the major competitors in green roof installations, which resulted in subsequent contracts.
- Market potential – what are the opportunities for expanding the company's client base?
    Can the company elevate its competitive status by acquiring the experience to perform such installations as green roofs, living walls, biofiltration ponds, and stream restoration?
    What are the capital requirements and potential return on investment?
- Value analysis – what is the project's profit potential?
- Government project compliance – does the company meet government project requirements for **MBE/WBE (Minority Business Enterprise/Women Business Enterprise)**
- Certifications, licenses – city/county
- Site review – accessibility, staging areas, existing conditions, utilities. What are the site conditions in relation to liability issues, e.g., crane requirements, traffic (urban site), employee safety?
- Risk factors – Is the project economically viable in terms of risk management costs such as bonding and liability insurance?
    - Can the company acquire a bond to meet the contract requirements?
    - Does the company have adequate resources, personnel, and equipment to mobilize the project?

- Can the company afford the financial risk associated with the project e.g., potential of being over budget; having 10% (of total contracted price) retainage fee?
- What are the per diem penalties (liquidated damages) for not meeting schedules? Potential delay factors include scheduling, weather, material shortages, equipment failures, and noise restrictions.
- Will the project have a negative impact on existing projects and its primary client base? This may be a factor both economically and physically in terms of reallocation of equipment and personnel. Some companies will actually put a cap on how large a project (in terms of dollars) that they will be willing to contract.

Planning/pre-engineering

Planning/pre-engineering is a comprehensive process (usually involving the client) that addresses all phases of the project in regard to establishing goals, evaluating engineering options, meeting specifications, and setting completion deadlines. It requires that the project manager and the production team address every detail, including allocating resources, establishing a comprehensive budget, setting goals and priorities, and establishing a firm timeline. Once this information has been compiled into a project database, field reports can then be generated for daily implementation. These reports will specify the amount of materials and production hours that are budgeted for the respective days. An accounting system is also established to track the project's progress on a daily basis.

Inputting the monitoring and control processes at this time will enable managers to discover issues that need to be addressed before the plan is implemented.

Monitoring and control

Monitoring and control processes are an integral part of project management that audits all the work being done on the project and provides verification for project managers as well as the client. Criteria are established to measure performance baselines, establish protocol for changes and for correcting defects, issue logs to monitor changes, establish communication channels for reporting performance, issue budget forecasts based on timelines and current progress, facilitate conflict resolution between construction teams and clients, and constantly monitor the technical specifications.

Execution

Execution starts with the establishment of open communication between the client and the general contractor, subcontractors, and landscape production teams through weekly

Project management

meetings. Status reports are presented for discussions on scheduling and anticipated delays due to material availability and weather conditions. **Lead time** is the amount of delay between placing an order and receiving the materials. Wet weather will delay digging nursery stock as well as installations. In addition, delays due to a short supply of hardscape or softscape materials may require recommendation of alternatives. Scheduling changes may be implemented due to subcontractor limited availability.

The project manager is responsible for the execution and control of the following:

- Scheduling (Figure 6.1) and maintaining production logs (Figure 6.2).
- Holding production meetings with contractors, subcontractors, clients/owners, architects, and engineers.
- Submitting samples of materials.
- Managing vendors – scheduling deliveries, processing invoices.
- Coordinating the staging area – trailer location, material unloading areas and holding areas, marking utilities.
- Job cost management (Figure 6.3) – current financial status, materials and labor, actual versus budget.
- Purchase order management – requires approval and signature.
- Billing – submission of monthly invoices to the general contractor, usually based upon percentage of completion.
- Change order management (Figure 6.4) – submission of quotes for additional work beyond the contracted project specifications is based on time and materials and securing authorization prior to implementation.
- Ensuring that all specifications are being met by the company's crew as well as subcontractors under their supervision (Figure 6.5).

Closure

Closure is the process that includes reviewing specification compliance, demobilizing the staging area, and conducting a project assessment:

- Site clean-up.
- Submission of **as-builts** (detailed plans that show in actuality how something was built) and maintenance manuals (mechanical and landscape management procedures).
- Inspection with general contractor, client, and landscape architect, verifying fulfillment of specifications.
- Formal acceptance by client.

SCHEDULE

		Start Date	Finish Date	# of Days	June 13-17	June 20-24	June 27-July 1	July 4-8	July 11-15	July 18-22	July 25-29	Aug 1-5	Aug 8-12	Aug 15-19
1	Contract award	6/15/09	6/15/09	1										
2	Notice to proceed	6/20/09	6/20/09	1										
3	Mobilization (trailer, hook-up, perimeter fence installation)	6/23/09	6/29/09	7										
4	MOE approval of sediment control drawings	6/30/09	6/30/09	1										
5	Erosion and sediment control measures installed	7/5/09	7/12/09	8										
6	Demolition of hardscape and plants (Phase 1)	7/13/09	8/19/09	39										
7	Demolition of hardscape (Phase 2)	10/31/09	11/15/09	16										
8	Lighting removal	7/13/09	7/13/09	1										
9	Import soil & spreading	7/21/09	9/21/09	44										
10	Survey & stake out	various	various	11										
11	Submittal submission (see attached submittal list)	8/1/09	8/12/09	12										
12	Tree tagging by landscape architect	8/8/09	8/12/09	5										
13	Storm drains and inlets	8/9/09	9/23/09	35										
14	Fire hydrant relocation	8/25/09	8/27/09	3										
15	Electrical sleeving & rough in	9/6/09	9/23/09	14										
16	Irrigation sleeving	9/9/09	9/14/09	5										
17	Crosswalk demolition and installation (preinker, drive street closure)	8/11/09	8/18/09	8										
18	Sub-grade inspection/testing	various	various	4										
19	Concrete sub-base for pavers	9/14/09	10/7/09	18										
20	Concrete curbs	8/25/09	9/15/09	16										
21	Brick nodes (Phase 1)	10/10/09	10/28/09	19										
22	Brick nodes (Phase 2)	11/21/09	12/9/09	19										
23	Concrete (Phase 2)	11/21/09	12/9/09	19										
24	Concrete walk w/brick bands	9/21/09	10/12/09	16										
25	Brick pavers	10/3/09	12/2/09	61										
26	Asphalt	10/17/09	10/28/09	12										
27	Base planting	10/17/09	11/30/09	44										
28	Planting substantial completion	12/1/09	12/1/09	1										
29	Turf	11/24/09	12/14/09	21										
30	Site furniture/ballard Installation	12/5/09	12/14/09	10										
31	Electric fixtures	12/21/09	01/14/10	24										
32	Railings	11/14/09	11/18/09	5										
33	Precast curb	10/24/09	11/4/09	12										
34	12" caliper trees	11/14/09	11/30/09	17										
35	Irrigation	10/3/09	12/9/09	68										
36	Plant maintenance until project completion	12/2/09	1/9/10	39										
37	Plant maintenance (All plants under 10"cal-1 year maint.Over10"cal-2year maint.)	1/9/10	1/9/11	730										
38	Erosion and sediment control removed	12/15/09	12/21/09	7										
39	Substantial completion	12/15/09	12/15/10	1										
40	Punch out	12/15/09	1/9/10	26										

Figure 6.1 Project Schedule

Source: Ruppert Landscape, Laytonsville, MD

PROJECT LOG
UMCP Southwest District Log

Date		On-Site	Subs	# of Emp.
14-Jul	Asphalt removal continued.		MJR	3
15-Jul	Asphalt removal and misc. hand work by MJR. Downloaded site info w/Drew. Heavy rain (at least 1").	JM & DD	MJR	5
16-Jul	Continued to remove asphalt. Removed majority of trees around perimeter. Began breaking up curbs.		MJR	4
17-Jul	Sunday.			
18-Jul	Continued to excavate parking lot. Removed +/- 20 dump loads. Continued curb demo. Tractor disabled. Benches secure.	DD	MJR	3
19-Jul	Continuing excavation of asphalt. Finished rest of trees. Started bringing in fill. Continued curb, brought close to trailer.	DD	MJR	3
20-Jul	P.M. #3 @ 10.00 continued last of asphalt parking. Finished removing curbs (still on site) Fire marshall came.	DD & RS	MJR	3
21-Jul	Asphalt Removed. Ballords removed and stored. +/-8 loads of import base soil dumped on site.	DD	MJR	3
22-Jul	Continued removing sub base. 5 loads of borrow dumped on site. Borrow Soil Approved by AMT. Started spreading.	DD	MJR	3
23-Jul	Continued moving out sub base. Took out sidewalk along northeast driveway of site. Moved curb stops 2 staging area.	DD	MJR	3
24-Jul	Sunday			
25-Jul	Hard rain in early morning. Unable to perform work. Looked over submittal information.	DD		
16-Jul	Disconnected and removed emergency phone. Continued on concrete & sub-base removal and importing soil.	JM & DD	MJR	3
27-Jul	Leveling off Soil. Removed more sub base and concrete. Electrician spliced light poles.	JM & DD	Nat/MJR	2+3
28-Jul	Continued spreading soil. Removed more sub base and concrete. Compaction started in upper corner.	JM	MJR	3
29-Jul	Spread and compact upper p-lot area, haul-off & import soil. Met w/AMT regarding survey needs. Heavy rain @ 2:30.	JM & DD	MJR	3
30-Jul	Saturday			
31-Jul	Sunday			
1-Aug	Soil compaction continued. Removed more sub base and Concrete. Surveyors began elevations and curb stake out.	JM & DD	MJR/AMT	3+2
2-Aug	More compaction. Surveyors continued curb stake outs, elevations and radii. Met with paver supplier.	JM & DD	MJR/AMT	3+2
3-Aug	P.M. #4 @ 10:30. Discussed pavement type for pedestrian drive. Concrete continued. More surveys.	JM & DD	MJR/AMT	2+2
4-Aug	Jim from university marked HVAC lines, water lines and storm drain. Mike from Trigen marked HV lines.	JM & DD	MJR	2
5-Aug	Mike from Trigen came back marked HV. John from Nationwide checked his work area. Waited for U of MD to mark utility.	JM & DD	TR/Nat	1+1
6-Aug	Saturday			
7-Aug	Sunday			

Figure 6.2 Project Log/Daily Reports
Source: Ruppert Landscape, Laytonsville, MD

JOB COST

Project Name			Material			Labor			Hours Per Unit	
Description	Size	U.M.	Est.	To Date	% Comp.	Est.	To Date	% Comp.	Est.	Actual
Materials										
Labor										
Landscape										
Hardscape										
Other										
Cleanup										
Watering										
Travel										
					Totals					

Figure 6.3 Job Cost Sheet

Source: *Blueprint for Success*, National Association of Landscape Professionals, Herndon, VA

CHANGE ORDER

Date ______________________ Job Number ______________________

Client Name ____

Address ______________________ City, State, Zip ______________________

Work Item	+/- Qty.	Unit of Measurement	Description	Size	Unit Cost	Total Price
					Total Contract Price Adjusted +/-	

Customer Authorization ______________________ Date ______________________

Figure 6.4 Change Order

Source: *Blueprint for Success*, National Association of Landscape Professionals, Herndon, VA

Project management

Figure 6.5 Ruppert Landscape Tree Installation, University of Maryland

- Demobilization – removal of all equipment, excess materials, trailer.
- Project assessment – evaluate production efficiency and conduct a procedural analysis; identify inefficiencies and address them by improving the respective procedures in which they occurred.
  Was the project profitable? Was it under budget (a buyout) or over budget (a bust)? Did it create new opportunities, e.g., maintenance contracts, establish new relationships with general contractors, provide experience in a new market niche?
- Document lessons learned and archive all production records.

Implementation/accountability

Team effort is vital to the successful execution of a project. Team leaders need to be organized, experienced, and capable of motivating and directing their team members. Their effectiveness is dependent upon their leadership skills. One of these skills is to be able to delegate without micromanaging. Team members are empowered to take responsibility and allowed to perform to the best of their ability. As a member of the team, these individuals meet with the project manager on a daily basis to assess their progress and on a weekly basis with the entire team. The weekly meetings are essential for the coordination of the team and keeping everyone updated as to the project status, e.g., budget, timeline, contingencies.

Using project tracking software enables instant communication of a project's financial status, which can be made accessible via the web on managers' tablets or smart phones. Daily job reports enable project managers to audit production efficiency and establish accountability for each team member based upon their performance. Daily audits held as an open forum with team leaders enables constructive comments to be shared on the current status of the project and upcoming phases. This enables managers to direct team leaders to make the necessary adjustments in their production phase to get back on track with the budget and ensure that quality standards and specifications are being met.

Measurement indices gauge the productive efficiency of the project. As previously stated, we are what we measure. Benchmarks (goals) are set for each phase of the project. Daily and weekly summaries can be compiled with actual labor hours versus budgeted to assess production efficiency. Since the budget hours were based on the company's production standards, this measurement provides an indication of the team's job performance. These measurements will raise red flags where the team needs to improve and whether additional equipment or personnel need to be allocated.

A project's chance of success is greatly enhanced if a company's standard procedures are properly executed. At the completion of each project, these standards are scrutinized for possible further optimization to maximum efficiency.

SUMMARY

What are best management practices?

Best management practices (BMP) are comprised of standardized project procedures that have been derived from previous projects. They represent procedures that have been proven to be the most efficient in implementing specific operations. The BMP are kept in a database in a modular form that enables them to be utilized in various phases of a construction operation regardless of its size. These management practices are analogous to the production standards that are established for landscape maintenance and installation operations.

What is sequential implementation of project management processes?

Sequential implementation of the project management processes is a structured approach to profiling and managing projects. The project manager and production team work together to

Project management

build a project foundation and its management infrastructure. The process begins with an assessment of the project's scope and ends with the project's closure.

The *scope assessment* component of the process involves a review of all aspects of the project to essentially determine whether a company has the expertise and resources to successfully manage the project. It also evaluates the potential profitably of the project.

The *planning/pre-engineering* process develops the project's management infrastructure. This entails the development of budgets, schedules, timelines, and a tracking system, as well as meeting protocols.

Monitoring and controlling processes provides audits for all phases of the project production. Criteria are established to measure performance baselines, develop protocol for changes and correcting defects, and ensure communication between construction teams and the client.

Execution is the implementation process that encompasses every activity occurring on the site. The project manager's average day includes managing vendors and subcontractors, reviewing specifications, meeting with clients, and keeping tabs on job cost management.

Closure of a project involves reviewing specifications for the entire project. The client and project manager walk the site and determine if any additional work is necessary to meet those specifications. Demobilization and site cleaning are scheduled. An assessment of the entire project is made by the project team and management that will determine where procedures could have been improved. It will also identify efficient procedures that will be incorporated into future best management practices.

What is project management accountability?

Project management accountability involves comparing actual production efficiency measurements to the projected allotments to keep the project on track. The project manager audits materials and labor costs through a job cost tracking system to be alert as to where adjustments need to be made. Budgeted hours are monitored on a daily basis. Daily inspections are conducted to monitor compliance with the specifications stipulated in the contract and establish accountability. Weekly meetings with team leaders and subcontractors ensure that production schedules will be current and that individuals are responsible for meeting their respective schedules. If scheduling deadlines are not met, the per diem penalty may be assessed.

KNOWLEDGE APPLICATION

1   Based on your knowledge of the landscape industry, list five operation procedures that you would consider to be best management practices. Describe your selection criteria.

Project management

2. In assessing the scope of a project, what aspects would make you consider it as having market potential for a landscape company?
3. List five risk factors associated with landscape construction projects and how you would mediate them.
4. As a project manager, what control measures would you implement to keep your project on schedule and on budget?

7

Financial ratios

CHAPTER OBJECTIVES

To gain an understanding of the components of the various categories of financial ratios, how they are calculated, what they indicate to financial management, and how to use financial ratios to maximize profitable operation of the company:

1. The categories of financial ratios
   a. Liquidity
      i. Current
      ii. Quick
      iii. Cash to current liabilities
   b. Debt
      i. Debt to equity
   c. Activity
      i. Inventory turnover
      ii. Average collection period
      iii. Accounts payable
   d. Productivity
      i. Sales per employee
      ii. Sales to fixed assets
   e. Profitability
      i. Gross margin
      ii. Profit margin
      iii. Return on assets
      iv. Return on equity

Financial ratios

2 The components of financial ratios
3 How ratios are calculated
4 Ratios as indices of:
   a Financial management
   b Productivity
   c Profitability
5 Improving ratios with astute financial management

KEY TERMS

accounts payable payout period	costs of goods sold	liquidity status
average collection period	current ratio	profitability ratios
average holding period	debt ratio	profit margin
cash to current liabilities ratio	gross margin	quick ratio
	inventory turnover ratio	return on assets (ROA)
	liquidity ratios	return on net worth

Ratios provide us with a perspective of relationships between two entities. They are found by dividing the magnitude of one entity into the other. We hear and read about ratios every day, particularly in sports. A college basketball point guard who averages twenty shots per game and makes five baskets has a ratio of baskets/attempted shots of 5/20 or 25%. This information assesses the player's shooting ability, which at 25% would not raise any eyebrows or attract any NBA offers.

Financial ratios also provide an index of performance. They indicate how well a business is financially managed. These ratios are calculated from information extrapolated from financial statements. Business ratios are categorized into liquidity ratios, debt ratios, activity ratios, productivity ratios, and profitability ratios.

LIQUIDITY RATIOS

The **liquidity status** of a company indicates its ability to pay short-term debts with current assets. Liquidity refers to the state of being liquid in terms of cash and assets that can be readily converted into cash. Personal liquidity status would include cash in savings and checking accounts, certificates of deposit, and stocks.

Liquidity ratios measure the current assets of a company against its current liabilities. *Current assets* are those that can be converted into cash within a twelve-month period. These

Financial ratios

assets include inventory, accounts receivable, cash, and securities, such as certificates of deposit, money market funds, and stocks. *Current liabilities* refer to debts that will be paid within a twelve-month period. These liabilities include loan payments, vendor invoices, state and federal taxes, payroll, insurance premiums, leases, and rent.

Three primary ratios measure a company's ability to meet its current financial obligations with its current assets. These ratios are the current ratio, quick ratio, and cash to current liabilities.

Current ratio

The amount of current assets and current liabilities is obtained from the company's balance sheet. **Current ratio** indicates a company's ability to pay short-term debt – insurance, vendor invoices, payroll, leases – with cash from savings and checking, accounts, securities, inventory, and accounts receivable. This is an important index for a company to monitor because of the seasonal fluctuations in the landscape industry. The impact of these fluctuations is evident in the annual cash flow pattern, which inevitably has periods of negative cash flow (expenses exceeding cash income).

$$\text{Current ratio} = \frac{\text{Current assets}}{\text{Current liabilities}}$$

A current ratio of one indicates that for every dollar of liabilities there is a dollar of assets. Therefore, a company with a ratio of one is able to meet its current liabilities or short-term debt with its current assets. Banks and loan institutions prefer a current ratio of two when reviewing a company's application for a loan or credit line, since it provides a greater safety margin with two dollars of current assets for every dollar of current liabilities.

Quick ratio

The **quick ratio** is sometimes referred to as the "acid-test ratio" because it indicates whether a business can pay its short-term debt with assets that can be converted into cash in the shortest amount of time, meaning readily accessible cash from bank accounts, marketable securities, and accounts receivable.

$$\text{Quick ratio} = \frac{\text{Cash} + \text{Accounts receivable}}{\text{Current liabilities}}$$

Inventory is not included in the calculation because of the time factor associated with its liquidity. A quick ratio of two is acceptable, indicating that there are $2.00 of liquid assets available to meet every $1.00 of current financial obligations.

Financial ratios

Cash to current liabilities ratio

Although companies do not maintain sufficient cash reserves to meet 100% of their liabilities, they do maintain a percentage based on their specific financial needs, for example payroll, vendor invoices, early payment discounts (usually within ten days of receipt), and loan and lease payments. The **cash to current liabilities ratio** measures a company's ability to meet its financial obligations with actual cash in bank accounts. A general guideline for this ratio would be 40%, or $.40 of cash for every $1.00 of liabilities.

$$\text{Cash to current liabilities ratio} = \frac{\text{Cash}}{\text{Current liabilities}}$$

In regard to this ratio, the opinion of the CFO of a leading landscape construction/maintenance company is that it is obsolete due to today's treasury management products. His company "sweeps" funds into investments to earn interest.

The primary difference between the three liquidity ratios – current ratio, quick ratio, and cash to current liabilities ratio – is the source of the assets that are used to pay the current debt, which determines the rapidity with which they can be converted into cash.

- The assets in the *current ratio* require the longest period due to the time required for liquidation of inventory and investments. A portion of the assets in the *quick ratio* are available immediately from cash in bank accounts and short-term investments, such as certificates of deposit, and the remaining assets are available as cash when payments are received from accounts receivable.
- All of the assets in the *cash to current liabilities ratio* are immediately available for debt payment, since they are already in cash-based checking and savings accounts.

The examples that follow will illustrate the significance of these ratios in measuring the company's ability to pay its short-term debts with the various current assets. The balance sheet that has been created for BradCo Landscape Company (Table 7.1) will provide the financial information for the ratio calculations, along with information derived from its income statement (Table 7.2).

$$\text{Current ratio} = \frac{\text{Current assets}}{\text{Current liabilities}}$$

BradCo Landscape Company

Current assets: $302,600 (from balance sheet)
Current liabilities: $92,300 (from balance sheet)

Financial ratios

$$\text{Current ratio} = \frac{\$302{,}600}{\$92{,}300} = 3.3$$

Industry current ratio averages for the various sectors are:

Design/build	1.6
Exterior maintenance	1.4
Exterior installation	1.2

Source: *2012 Operating Cost Study for the Green Industry*, National Association of Landscape Professionals, Herndon, VA

$$\text{Quick ratio} = \frac{\text{Cash} + \text{Accounts Receivable}}{\text{Current liablities}}$$

BradCo Landscape Company

Cash: $16,300 (from balance sheet)
Accounts receivable: $132,500 (from balance sheet)
Current liabilities: $92,300 (from balance sheet)

$$\text{Quick ratio} = \frac{\$16{,}300 + \$132{,}500}{\$92{,}300} = \frac{\$148{,}800}{\$92{,}300} = 1.6$$

Industry quick ratio averages for the various sectors are:

Design/build	1.1
Exterior maintenance	1.2
Exterior installation	1.0

Source: *2012 Operating Cost Study for the Green Industry*, National Association of Landscape Professionals, Herndon, VA

$$\text{Cash to current liabilities ratio} = \frac{\text{Cash}}{\text{Current liabilities}} \times 100$$

BradCo Landscape Company

Cash: $16,300 (from balance sheet)
Current liabilities: $92,300 (from balance sheet)

$$\text{Cash to curent liabilities ratio} = \frac{\$16{,}300}{\$92{,}300} = 0.18 \times 100 = 18\%$$

Industry cash to current liabilities ratio averages for the various sectors are:

Design/build	34.5%
Exterior maintenance	25.9%
Exterior installation	10.6%

Source: *2012 Operating Cost Study for the Green Industry*, National Association of Landscape Professionals, Herndon, VA

So what do these ratios say about the performance of the BradCo Landscape Company?

The *current ratio* indicates that there is $3.30 of current assets for every $1.00 of current liability. Another way of looking at it is that the current assets are 330% of the current liabilities. Regardless of the way it is stated, BradCo Landscape Company has more than sufficient assets that can be converted into cash to meet its twelve-month liabilities.

The *quick ratio* indicates that there is $1.60 of cash and accounts receivable assets for every $1.00 of current liabilities. In other words, the twelve-month liabilities could be paid with the company's cash holdings, securities, and accounts receivable. Keep in mind, however, that the ratio is based on the accounts receivable being current (within thirty to forty days). Therefore, the company's credit department needs to be diligent in managing the accounts receivable to maintain a favorable quick ratio.

The *cash to current liabilities ratio* shows that there is only $.18 of cash assets for every $1.00 of short-term debt, or stated in percentage terms, cash assets can meet only 18% of the short-term debt. Cash holdings for the BradCo Landscape Company would be inadequate during periods of negative cash flow and therefore a bigger cash reserve should be maintained to better position the company to be able to meet a larger percentage of its short-term debt. A recommendation would be for the company to reduce its inventory and reallocate the invested dollars into cash accounts or short-term securities. Purchases of inventory in anticipation of allocation to jobs, as in this instance, ties up liquid capital that could be applied to current liabilities.

Taking all of the liquidity ratios into consideration, it is evident that current assets in total can meet the company's short-term obligations, but it would require liquidation of most of the current assets. A recommendation for the BradCo Landscape Company would be to build up a larger cash reserve and to be diligent in keeping accounts receivable current.

DEBT RATIO

A **debt ratio** measures the amount of a company's debt relative to the owner's equity (amount of money remaining after liabilities are subtracted from assets). It indicates the amount of debt that can be financed by the owner's equity. Companies with debt ratios larger than one

Financial ratios

would deter lending institutions from approving a line of credit or loan. Any ratio above one would indicate that the company's debt exceeds the owner's equity (net worth).

$$\text{Debt ratio} = \frac{\text{Total liabilities}}{\text{Owner's equity}}$$

BradCo Landscape Company

Total liabilities: $398,800 (from balance sheet)
Owner's equity: $445,180 (from balance sheet)

$$\text{Debt ratio} = \frac{\$398,800}{\$447,180} = .89$$

Industry debt to equity ratio averages are:

Design/build	.7
Exterior maintenance	.7
Exterior installation	1.8

Source: *2012 Operating Study for the Green Industry*, Professional Landcare Network, Herndon, VA

BradCo's debt ratio of .89 indicates that there is $.89 of debt for every $1.00 of owner's equity. The company has less debt than equity and is therefore able to fund its business operations without the need for outside financing. Any outside financing will burden the company with additional liabilities associated with the principal and interest of a loan.

ACTIVITY RATIOS

Two business entities that contribute to cash flow are inventory and accounts receivable. Ratios that indicate their activity – conversion into cash – are useful tools for monitoring the period of time it takes for the conversion.

The income statement (Table 7.2), along with the balance statement in Table 7.1, provide financial information for the activity, productivity, and profitability calculations that follow.

Inventory turnover ratio

An **inventory turnover ratio** (ITR) indicates the frequency with which inventory is converted into cash via sales to jobs. Whether the inventory is fieldstone or maple trees, inventory doesn't

Table 7.1 A Balance Sheet Summarizes a Company's Assets and Liabilities and Reflects Its Net Worth (Owner's Equity)

BradCo Landscaping Company
Balance Sheet
12/2016

Assets
Current assets
Cash	$16,300	
Securities	7,000	
Accounts receivable	132,500	
Inventory	146,800	
Total current assets	$302,600	

Fixed assets
Building/improvements	$360,000	
Equipment and vehicles	158,800	
Furniture and fixtures	50,100	
Depreciation allowance	−25,620	
Total fixed assets	$543,380	
Total assets		$845,980

Liabilities and owner's equity
Current liabilities
Accounts payable	$37,000	
Accrued expenses	27,600	
Taxes payable	3,500	
Notes payable	24,200	
Total current liabilities	$92,300	

Long-term liabilities
Notes payable	$306,500	
Total liabilities		$398,800

Owner's equity
Stock holdings	$212,000	
Retained earnings	235,180	
Total owner's equity (net worth)		$447,180
Total liabilities and equity		$845,000

Financial ratios

Table 7.2 A Company's Income Statement Indicates the Current Profitability of Its Operations for the Respective Statement

BradCo Landscaping Company
Income Statement
05/2017

Net sales	$1,370,000
Direct costs	
Labor	$383,600
Materials	205,500
Subcontractors	82,200
Total direct costs	$671,300
Gross margin	$698,700
Indirect overhead	$191,800
G & A overhead	$356,200
Pretax profit	$150,700

provide any sales revenue to the company until it is sold to a job. Management of inventory (inventory control) requires knowledge of the quantity of materials needed and the time frame. Inventory that sits in the yard for long periods costs the company money. In regard to plant materials, there are costs associated with maintenance and losses due to weather factors.

In order for this ratio to accurately reflect inventory turnover, every item used for a job needs to be assigned a job number. This process ensures that the cost of inventory will be recovered as a direct cost to jobs. Monitoring inventory outflow is an integral part of an inventory control system. The ITR is calculated by dividing the wholesale cost of goods sold (materials) by the average inventory within a specific period.

$$\text{Inventory turnover ratio} = \frac{\text{Cost of good sold}}{\text{Average inventory}}$$

BradCo Landscape Company

Materials direct costs: $205,500 (from income statement)
Average monthly inventory: 148,000 ÷ 12 = $12,333

$$\text{Inventory turnover ratio} = \frac{\$205,00}{\$12,333} = 16.66$$

Financial ratios

Inventory average turnover rates for landscape contracting businesses are:

Exterior design/build	12.0
Exterior maintenance	16.8
Exterior installation	14.4

Source: *Source 2012 Operating Cost Study*, National Association of Landscape Professionals, Herndon, VA

These turnover rates indicate the number of times the dollar value of the inventory is sold on an annual basis. BradCo's inventory turnover rate of 16.66 indicates that its average inventory is converted into cash approximately 17 times a year, for an **average holding period** of 22 days (365/16.66) This turnover rate is within the industry benchmark for the maintenance sector of the industry.

Average collection period

Accounts receivable has an even more significant impact on cash flow because of the dollars involved. Accounts receivable turnover is expressed as an **average collection period**. This financial indicator reflects the period of time, in days, that it takes a company to collect its accounts receivable (credit sales). This is found by comparing the accounts receivable to average daily credit sales. Average daily credit sales is calculated by dividing net credit sales by the number of days in the sale period:

$$\text{Average daily credit sales} = \frac{\text{Net credit sales per period}}{\text{Days in period}}$$

$$\textbf{Average collection period} = \frac{\text{Accounts receivable}}{\text{Average daily credit sales}}$$

BradCo Landscape Company

Accounts receivable: $132,500 (from balance sheet)
Net sales: $1,370,000 (annually, from income statement)
Days in period: 365

$$\text{Average daily credit sales} = \frac{\$1,370,000}{365} = \$3,753$$

$$\text{Average collection period} = \frac{\$132,500}{\$3,753} = 35.3$$

Financial ratios

Average collection days for various sectors of the landscape industry are:

Exterior design/build	24.8
Exterior maintenance	28.0
Exterior installation	34.8

Source: *2012 Operating Cost Study for the Green Industry*, National Association of Landscape Professionals, Herndon, VA

Since accounts receivable constitute the primary cash flow entity of landscape companies, their accounting departments need to be diligent in invoicing and collecting payments. BradCo's average collection period falls within the good range of thirty to forty days.

Accounts payable payout period

Another business activity that warrants close monitoring is accounts payable. The period of time it takes a company to pay its vendors (suppliers) is indicative of a company's financial management. Vendors will check a company's **accounts payable payout period** prior to extending credit. This information can be derived from business subscription services such as Dun and Bradstreet. Paydex is a term used in business, for a numerical score granted by Dun and Bradstreet to business as a credit score for the promptness of their payments to creditors.

The number of days a company takes to pay its vendors for a specific period is calculated by the formula:

$$\text{Accounts payable payout period} = \frac{\text{Average accounts payable}}{\text{Average daily costs of goods sold}/365}$$

Average accounts payable for a specific period can be found by adding beginning accounts payable for the period to the ending accounts payable and dividing by two. The average daily **costs of goods sold** by the company refers to the direct costs for materials as found on the income statement.

BradCo Landscape Company

Accounts payable: $37,000 (from balance sheet)
Materials direct costs (annual): $205,500 (from income statement)
Days in period: 365

$$\text{Average daily costs of goods sold} = \frac{\$205{,}500}{365 \text{ days}} = \$563 \text{ per day}$$

Financial ratios

$$\text{Accounts payable payout period} = \frac{\$37,000}{\$563} = 65.7 \text{ days}$$

Industry average days for accounts payable period for the various sectors:

Exterior design/build	43.6
Exterior maintenance	78.1
Exterior installation	49.0

Source: *2012 Operating Cost Study for the Green Industry*, Professional Landcare Network, Herndon, VA

Vendors will grant credit without reservation to companies that pay their invoices within thirty to forty days. If the timeliness of payments exceeds sixty days (as BradCo's does), vendors may require C.O.D. (cash on delivery) for materials/supplies. In the BradCo Landscape Company example, all of their vendors will either require C.O.D. on all orders until delinquent invoices are paid or assess an interest penalty on the invoice balance. These delinquent invoices may be associated with inadequate cash reserves due to excessive inventory, delinquent accounts receivable, and/or negative cash flow associated with a large installation project(s). A new company normally will have to do business with a vendor for several months to a year to establish an acceptable payable period (thirty to forty days) before credit will be extended.

PRODUCTIVITY RATIOS

Two primary ratios reflect the productivity of a company. One relates to employees, and the other relates to fixed assets. They are both indicators of the efficient use of these production resources.

Labor costs

Labor is the largest cost category on a landscape company's income statement. It is the goal of every company to maximize the production from its labor force. As productivity increases, the return on labor costs will be reflected in the ratio of sales per full-time employee (FTE). Seasonal employees are also allocated into the full-time equivalents on the basis of two seasonal employees being equivalent to one full-time employee.

$$\text{Sales per employee} = \frac{\text{Net sales}}{\text{Number of full-time equivalent employees}}$$

Financial ratios

BradCo Landscape Company

Net sales: $1,370,000 (from income statement)
Full-time equivalent employees: 25 (from personnel records)

$$\text{Sales per employee} = \frac{\$1{,}370{,}000}{25} = \$54{,}800$$

The employee sales ratio of $54,800 is indicative of the amount of production dollars that can be generated by BradCo Landscape's full-time employees working in their profit centers.

Industry averages for sales per employee among the various sectors are:

Exterior design/build	$82,860
Exterior maintenance	$62,185
Exterior installation	$78,668

Source: *2012 Operating Cost Study for the Green Industry,* National Association of Landscape Professionals Herndon, VA

In addition to indicating the sales dollars of FTEs, this ratio also assists the company in planning for growth. Based on the sales productivity of employees, management can project the number of employees needed to meet a projected sales figure. If the sales productivity per employee is not being maintained annually, technical training, employee motivation, and equipment assessment may be required. Although the benchmarks below reflect industry averages, the benchmarks within a company are management's benchmarks. By analyzing company and division productivity ratios, management can identify which factors – equipment, scheduling, or training – will increase production efficiency.

Fixed assets

Equipment and vehicles make up the largest component of a landscape company's fixed assets. The return on this capital investment is based on the amount of production (sales) that it generates. The productivity ratio of **sales to fixed assets** represents the percentage of return from the dollars invested in fixed assets. The productivity ratio for fixed assets is found with the following calculation:

$$\text{Sales to fixed assets} = \frac{\text{Net sales}}{\text{Net fixed assets}}$$

Financial ratios

BradCo Landscape Company

Net sales: $1,370,000 (from income statement)
Total fixed assets: $543,380 (from balance sheet)

$$\text{Sales to fixed assets} = \frac{\$1,370,000}{\$543,380} = 2.5$$

Industry averages for sales to fixed assets among the various sectors are:

Exterior design/build	12.5
Exterior maintenance	10.6
Exterior installation	8.3

Source: *2012 Operating Cost Study for the Green Industry*, National Association of Landscape Professionals, Herndon, VA

The 2.5 ratio in the example indicates that for every $1.00 of BradCo Landscape Company's fixed assets, $2.50 in sales is generated. In this example, BradCo Landscape Company's low ratio reflects a recent investment in buildings and improvements for a new branch location ($360,000). As additional sales are generated from this branch, the percentage return from fixed assets will increase.

PROFITABILITY RATIOS

Additional indices of a company's financial performance are **profitability ratios**. The profitability of a company is measured in relation to sales, total assets, and net worth. **Gross margin** measures profitability after direct costs associated with production are subtracted from sales.

$$\text{Gross margin} = \frac{\text{Gross profit dollars}}{\text{Net sales}} \times 100$$

BradCo Landscape Company

Gross margin: $698,700 (from income statement)
Net sales: $1,370,000 (from income statement)

$$\text{Gross margin} = \frac{\$698,700}{\$1,370,000} \times 100 = 51\%$$

Financial ratios

The average gross margin for the various industry sectors is:

Exterior design/build	44.9%
Exterior maintenance	51.7%
Exterior installation	48.7%

Source: *2012 Operating Cost Study for the Green Industry*, National Association of Landscape Professionals, Herndon, VA

The more efficient a company is in its production of services, the larger the gross margin will be as a result of reduced costs. Based on financial history, the gross margin benchmark must be met to attain projected net profits, assuming overhead and administrative expenses remain relatively constant.

Profit margin measures the percentage of profit after direct costs and overhead are subtracted from sales. It provides another index with which companies can assess their financial performance against competitors. Trade association surveys provide benchmarks that are representative of specific sectors within the landscape industry.

$$\text{Profit margin} = \frac{\text{Pretax profit}}{\text{Net sales}} \times 100$$

BradCo Landscape Company

Pretax profit: $150,700 (from income statement)
Net sales: $1,370,000 (from income statement)

$$\text{Profit margin} = \frac{\$150,700}{\$1,370,000} \times 100 = 11\%$$

The average profit margin percentage for the industry sectors are:

Exterior design/build	1.5%
Exterior maintenance	9.2%
Exterior installation	9.2%

Source: *2012 Operating Cost Study for the Green Industry*, National Association of Landscape Professionals, Herndon, VA

The bottom line is profit, which is the primary indicator of a company's production efficiency. Production efficiency is the primary factor associated with high-profit companies. Profit is the fuel for the company's growth engines; the larger the fuel supply, the faster the

Financial ratios

engines will drive the company ahead of its competition. BradCo's 11% net profit indicates both production efficiency and astute overhead cost management.

Two additional measures of profitability are **return on assets (ROA)** and **return on net worth** (owner's equity). These two ratios are of particular interest to owners and stockholders. They both reflect the percentage return on dollars that are invested into the business.

The return on assets measures the company's profit as a percentage of the company's assets.

$$\text{Return on assets} = \frac{\text{Pretax profit}}{\text{Total assets}} \times 100$$

BradCo Landscape Company

Pretax profit: $150,700 (from income statement)
Total assets: $845,980 (from balance sheet)

$$\text{Return on assets} = \frac{\$150,700}{\$845,980} \times 100 = 18\%$$

The BradCo Landscape Company's ROA indicates that for every dollar it has invested in total assets (current and fixed) it receives a return of .18, or 18%. The profitability of the remainder of the year will determine whether this percentage of return prevails. Contributing to this rate of return is management of production costs and capital investments in equipment. In regard to the latter, BradCo rents most of the large equipment required for its installation projects, thereby eliminating the massive expenditure of purchasing heavy equipment and bearing the expense of using heavy equipment only as needed.

The industry averages for return on assets are:

Exterior design/build	6.1%
Exterior maintenance	8.5%
Exterior installation	10.4%

Source: *2012 Operating Cost Study for the Green Industry*, National Association of Landscape Professionals, Herndon, VA

The ROA indicates the amount of profit that is being generated by the company's total assets. The exterior maintenance contractor ROA average of 8.5% indicates that the average return for those companies is $.085 for every $1.00 of company assets. A business owner would be satisfied with any percentage that exceeds what the same asset dollars could return in banks or securities.

Financial ratios

The return on net worth measures the percentage of profit based on the company's net worth.

$$\text{Return on net worth} = \frac{\text{Pretax profit}}{\text{Net worth}} \times 100$$

It may also be expressed as:

$$\text{Return on equity} = \frac{\text{Pretax profit}}{\text{Owner's equity}} \times 100$$

The terms *company's net worth* and *owner's equity* are synonymous, in that they both represent the amount of money invested in the company after all liabilities have been met. Think of it in the same way as equity that remains in a house. Home equity represents the amount of money invested in the house as the difference between the mortgage balance (liability) and the value of the house (total revenues). The return on owner's equity/company net worth indicates what these invested dollars are earning based on the profitability of the company. Regardless of a favorable ROA, this does not merit complacency, since the management of assets, such as accounts receivable as well as profit margins, can always be improved and thereby increase the percentage of return on equity.

BradCo Landscape Company

Pretax profit: $150,700 (from income statement)
Owner's equity: $447,180 (from balance sheet)

$$\text{Return on equity} = \frac{\$150,700}{\$447,180} \times 100 = 33.7\%$$

The owners of the BradCo Landscape Company are receiving 33.7% return from their equity holdings in the company. This is a very good return in contrast to alternative investment options. In addition, this return is a positive reflection of management's ability to control costs and thereby produce a favorable level of profitability.

Industry sector averages for return on net worth/owner's equity are:

Design/build	15.9
Exterior maintenance	19.5
Exterior installation	32.2

Source: *2012 Operating Cost Study for the Green Industry*, National Association of Landscape Professionals, Herndon, VA

The *return on assets* and *return on net worth/owner's equity* ratios are measures of performance in relation to investment. The former reflects the return on asset investment and the

Financial ratios

latter reflects the return on every dollar invested by the owners. Both of the investment returns ratios are affected by two profitability ratios, gross margin and profit margin. Each of these ratios represents opportunities through astute financial management to increase the return on a company's investment. The gross margin is the end result of direct cost management, and profit margin is the end result of total cost and expense management – direct, indirect, and overhead.

Although additional ratios are used by financial analysts to assess company performance, those discussed in this chapter represent the primary financial assessment ratios. The ratios are additional tools for the financial toolbox. They assist management in measuring where the company is in relation to industry benchmarks and the company's financial history.

Financial analysis begins where accounting statements end.[1]

SUMMARY

Financial ratios are measurement tools. They provide a means of assessing the company's financial strengths and weaknesses as well as the performance of management.

What are the categories and components of financial ratios?

- **Liquidity ratios** measure the assets of a company against its liabilities. They indicate a company's ability to meet financial obligations with its assets. The ratios include the *current ratio*, *quick ratio*, and *cash to current liabilities ratio*. The computation of these ratios varies based on the liquidity of the assets used in their calculation.
- The **debt ratio** measures a company's debt relative to owner's equity. It provides an index of the amount of debt that can be financed by the owner's equity.
- **Activity ratios** measure the period of time required to convert inventory and accounts receivable into cash and the period of time that vendor invoices are paid. *Inventory turnover* measures the frequency with which the total dollar inventory value is converted into cash as sales to jobs. The *average collection period* measures the average amount of days required to collect accounts receivable. The *accounts payable payout period* measures the number of days a company takes to pay its suppliers.
- **Productivity ratios** measure the production of personnel and assets. *Sales per employee* are an index of individual productivity based on dollars of sale produced in relation to the number of full-time employees. *Sales to fixed assets* measures the amount of production that is the result of fixed asset allocation. It is an index of how effectively fixed assets are utilized.

Financial ratios

- **Profitability ratios** measure the profitability of sales and invested dollars from assets and owner's equity (net worth). *Gross margin* is an index of the profit generated after direct costs have been deducted from sales. *Profit margin* measures the profitability of sales income after direct costs and overhead costs have been deducted. *Return on assets* (ROA) reflects the profit generated by the company's total assets. *Return on equity* indicates the profit produced from an owner's invested dollars.

Why are financial ratios indices of business performance?

Each of these ratios measures a different aspect of the company's performance and also serves as a report card of how well management is doing.

How can management improve the ratios?

Employing these financial tools provides management with an opportunity to manage and adjust factors that will improve production efficiency and profitability.

KNOWLEDGE APPLICATION

1   Calculate the following ratios from the financial data provided:

Todd & Son Landscape Company

Total sales	$2,544,500
Cash	61,000
Inventory	185,000
Securities	6,000
Accounts receivable	269,000
Accounts payable	9,360
Current assets	256,020
Current liabilities	75,300
Total liabilities	619,900
Net worth	409,700
Materials direct costs	85,500

a   Current ratio
b   Quick ratio
c   Debt ratio

- d Average collection period
- e Inventory turnover ratio
- f Accounts payable payout period

2 What is your financial assessment of the calculated ratios in reference to:

- a The ability of the company to meet its financial obligations?
- b The company's ability to pay its debt with its net worth?
- c Timeliness in accounts receivable collections?
- d Inventory turnover rate?
- e Payable payment period?

3 What would be your goal for inventory turnover? How would you manage inventory to achieve this goal?

4 The management of Todd & Son Landscape Company has increased its gross margin from 35% to 45%, but its net profit remains at 4%.

- a What can be attributed to the increase in gross margin?
- b Why hasn't net profit increased with the increased gross margin?

NOTE

1 Anonymous, personal communication.

8

Software applications

CHAPTER OBJECTIVES

To gain an understanding of the wide variety of software applications used in the landscape industry.

1 Accounting systems
2 Estimating software
3 Smartphone applications
4 Design software
5 Mobile resource management systems
6 GIS imaging
7 Drone diagnostic imaging

KEY TERMS

global navigation satellite system (GNSS)
global positioning system (GPS)
inertial measurement unit (IMU)
light detection and ranging (LiDAR)
mobile resource management (MRM)

Computer technology has revolutionized the business world. Technology that initially facilitated data processing and information storage is now employed to service customers and provide integrated business management systems. Think of how often your daily life interfaces with computers just from the standpoint of meeting your consumer needs. Electronic banking lets you pay your bills; automated teller machines give you 24/7 access to cash; computerized ticketing machines at the airport or movie theater help avoid long

Software applications

lines; the automated self-service aisle in the grocery store talks you through the process of scanning your items, then processes your payment. Students are able to register for next semester's classes at their computer. E-business enables us to shop for cars, gifts, vacations, mortgages, and every other conceivable commodity, purchase our selection electronically, and schedule home delivery.

Efficient and expedient technology has changed the way we conduct our personal business and the way businesses market and sell their products and services. Businesses have substantially lowered their overhead by reducing the number of personnel required to generate and process sales. The technology has also enabled permanent records to be established for each transaction that reflect the purchasing patterns and demographics of the customer base. Although landscape companies can't deliver their services with computer technology, they can benefit from the expediency and efficiency associated with computerized information processing. Its implementation has enabled companies to access their current financial status and proactively manage production operations in real time to increase their profitability.

There are numerous software programs that have applications for the landscape industry. These programs are used to aid in:

- Accounting and budgeting
- Estimating, sales proposals, and contract and sales management
- Purchasing, inventory control, and job cost management
- Remote data collection
- Mobile resource management, scheduling, and routing
- Design development
- Irrigation system design and management
- Customer relation management
- Payroll and human resource management

Management needs to know the daily financial status of the company's branches and divisions. Management decisions regarding information technology (IT) implementation revolve around what information needs to be accessed on a daily basis. Instant access to current job cost information enables timely detection of production inefficiencies. Analysis of actual versus budgeted costs reveals production overages and enables adjustments to be made prior to job completion. Bruce Hunt, former vice president of the Brickman Group, Ltd. (currently BrightView), relates his company's philosophy and further states that corporations who do not abide by this philosophy will encounter financial failures.

With passion we strive to understand and measure where we are.[1]

Software applications

Knowing where you are enables you to make adjustments that will redirect you to where you want to be. These adjustments cannot be made without access to current data that is generated from financial reports.

Information technology provides the infrastructure for information management systems. The management aspect pertains to categorizing and interfacing information that is subsequently incorporated into report formats.

Many software accounting/management systems are available in today's market. The decision on which program(s) to implement in a company should be made by the accountant and the users: executives, managers, accounting staff, and other individuals such as estimators who input or access data. Some of the criteria that may be used to assist in the selection include:

- Interface between accounting operations
- Current and future needs
- Availability of technical support
- User friendliness
- Report capabilities
- Field applications
- Client invoicing
- Expandable with growth

Examples of information management systems available to the landscape contracting industry are presented in the following discussion.

ACCOUNTING SYSTEMS

This is the starting point for developing an integrated management system. The use of accounting software efficiently integrates financial information from various input sources for generating financial reports. The database in this system is the hub from which management extrapolates financial information for its day-to-day decisions. Budgets are updated on the foundation of current financial information, such as contract sales and job backlogs. Account managers are able to monitor their accounts' status by comparing the actual versus budgeted production hours and making adjustments. Branch managers are able to track each account manager's accounts and raise a red flag when necessary. Profit and loss statements reflect whether the company and individual divisions are meeting their projected profit margins.

Software applications

Companies need to adapt existing programs to their accounting system. It may take a year of working with consultants or technical support personnel before the system works to the satisfaction of the accounting department and management personnel. Some companies retain consultants to customize their own program. Regardless of its origin, the common goal for all accounting systems should be *integration*. The value of an integrated accounting system is its ability to interface with estimating, budgeting, and job cost management. Users must be able to extrapolate data, such as production rates, from the system and integrate the data into desired formats, such as job estimates. Simply stated, a company's software system should be able to manage all aspects of their business. An example of such a system is Aspire™ (https://youraspire.com), which was developed for the landscape contracting industry. It is an integrated system that develops a financial history from proposal to invoicing. Among its extensive reporting features is the real-time job cost reporting that enables managers to be proactive. Since it is Cloud based, its information is accessible from any mobile device.

The reports in the screenshots below are indicative of some of its managerial capabilities:

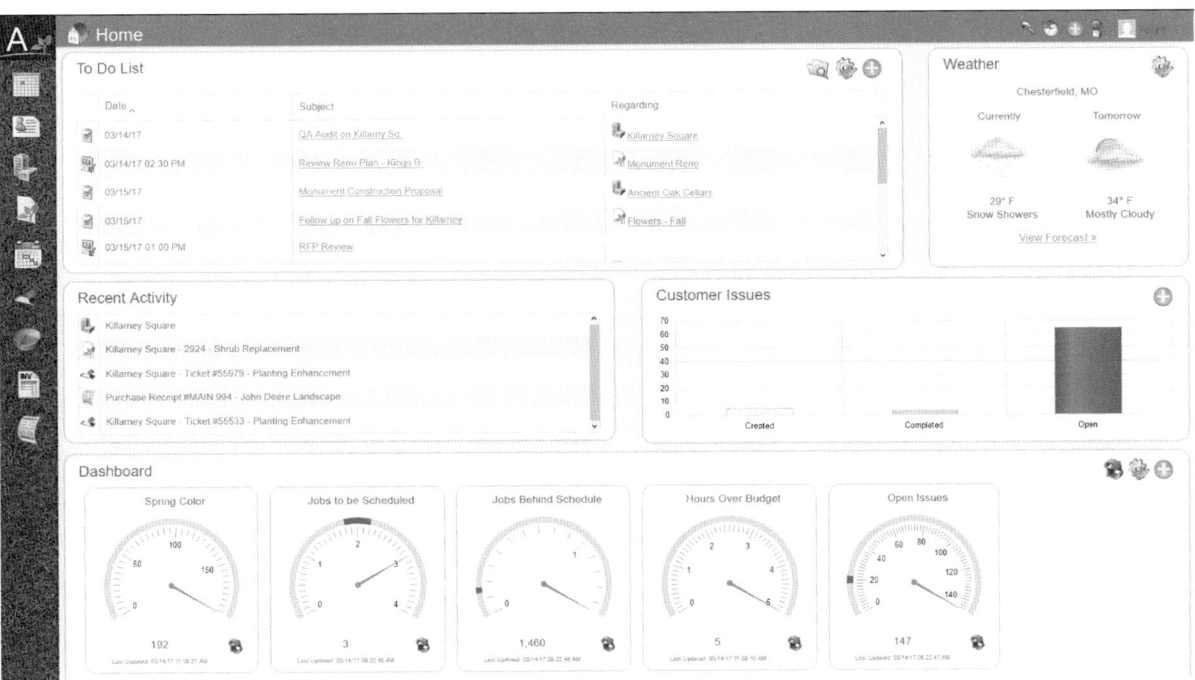

Figure 8.1 Dashboard

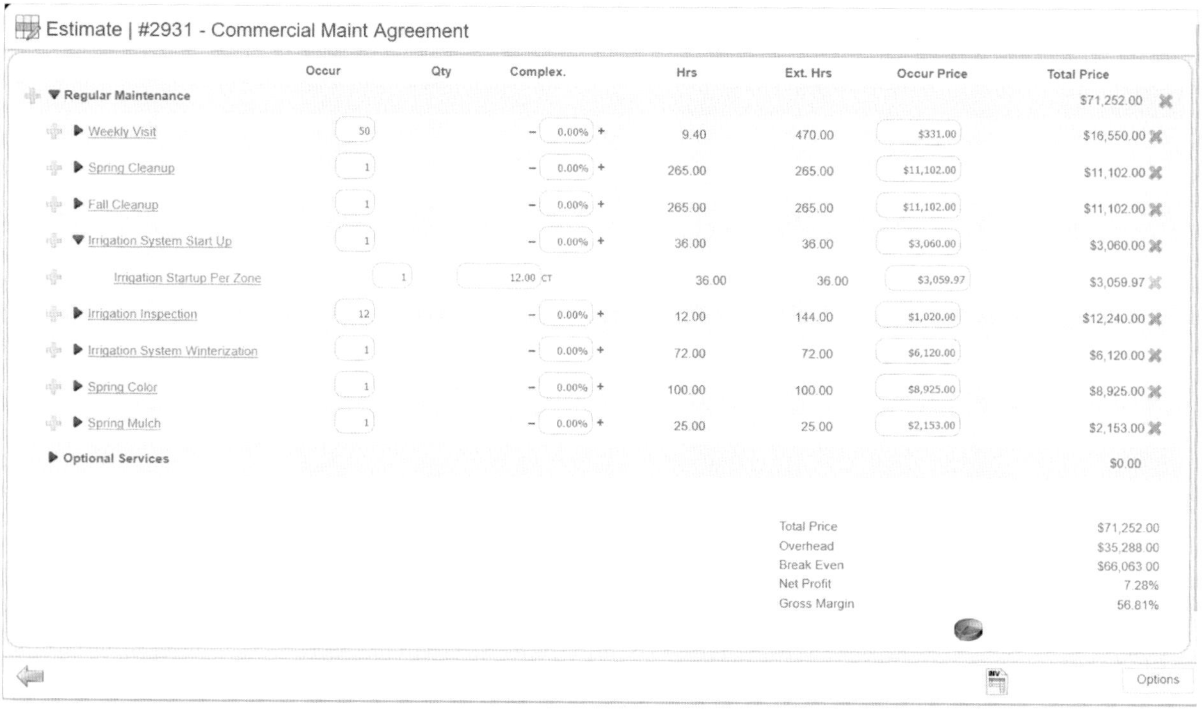

Estimate | #2931 - Commercial Maint Agreement

	Occur	Qty	Complex.	Hrs	Ext. Hrs	Occur Price	Total Price
Regular Maintenance							$71,252.00
Weekly Visit	50		0.00%	9.40	470.00	$331.00	$16,550.00
Spring Cleanup	1		0.00%	265.00	265.00	$11,102.00	$11,102.00
Fall Cleanup	1		0.00%	265.00	265.00	$11,102.00	$11,102.00
Irrigation System Start Up	1		0.00%	36.00	36.00	$3,060.00	$3,060.00
Irrigation Startup Per Zone		1	12.00 CT	36.00	36.00	$3,059.97	$3,059.97
Irrigation Inspection	12		0.00%	12.00	144.00	$1,020.00	$12,240.00
Irrigation System Winterization	1		0.00%	72.00	72.00	$6,120.00	$6,120.00
Spring Color	1		0.00%	100.00	100.00	$8,925.00	$8,925.00
Spring Mulch	1		0.00%	25.00	25.00	$2,153.00	$2,153.00
Optional Services							$0.00

Total Price	$71,252.00
Overhead	$35,288.00
Break Even	$66,063.00
Net Profit	7.28%
Gross Margin	56.81%

Options

Figure 8.2 Estimating Services

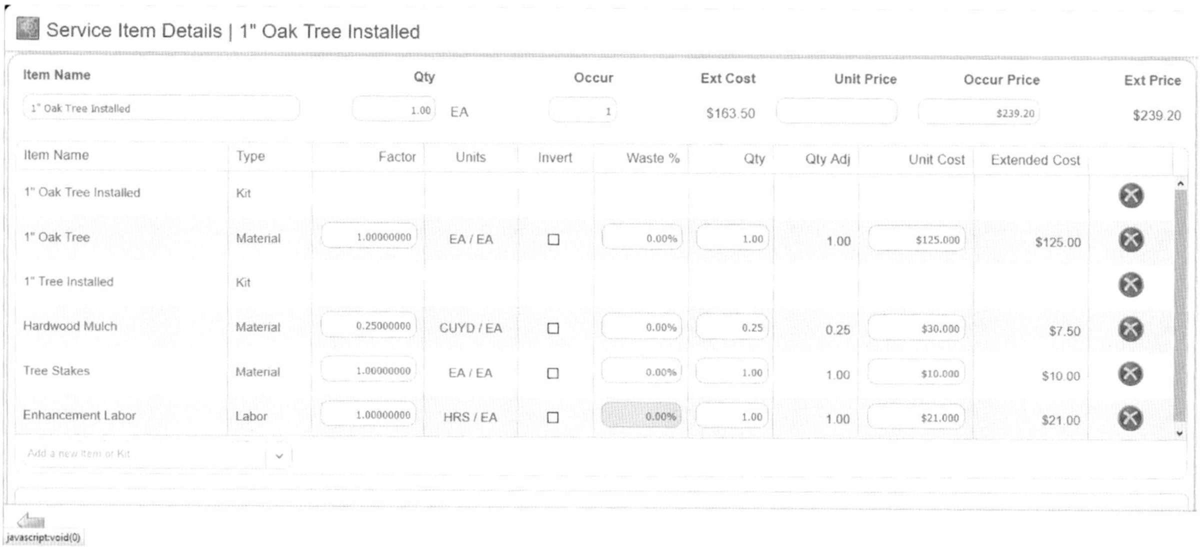

Service Item Details | 1" Oak Tree Installed

Item Name	Qty		Occur	Ext Cost	Unit Price	Occur Price	Ext Price
1" Oak Tree Installed	1.00	EA	1	$163.50		$239.20	$239.20

Item Name	Type	Factor	Units	Invert	Waste %	Qty	Qty Adj	Unit Cost	Extended Cost
1" Oak Tree Installed	Kit								
1" Oak Tree	Material	1.00000000	EA / EA	☐	0.00%	1.00	1.00	$125.000	$125.00
1" Tree Installed	Kit								
Hardwood Mulch	Material	0.25000000	CUYD / EA	☐	0.00%	0.25	0.25	$30.000	$7.50
Tree Stakes	Material	1.00000000	EA / EA	☐	0.00%	1.00	1.00	$10.000	$10.00
Enhancement Labor	Labor	1.00000000	HRS / EA	☐	0.00%	1.00	1.00	$21.000	$21.00

Add a new Item or Kit

javascript:void(0)

Figure 8.3 Estimating Tree Assembly

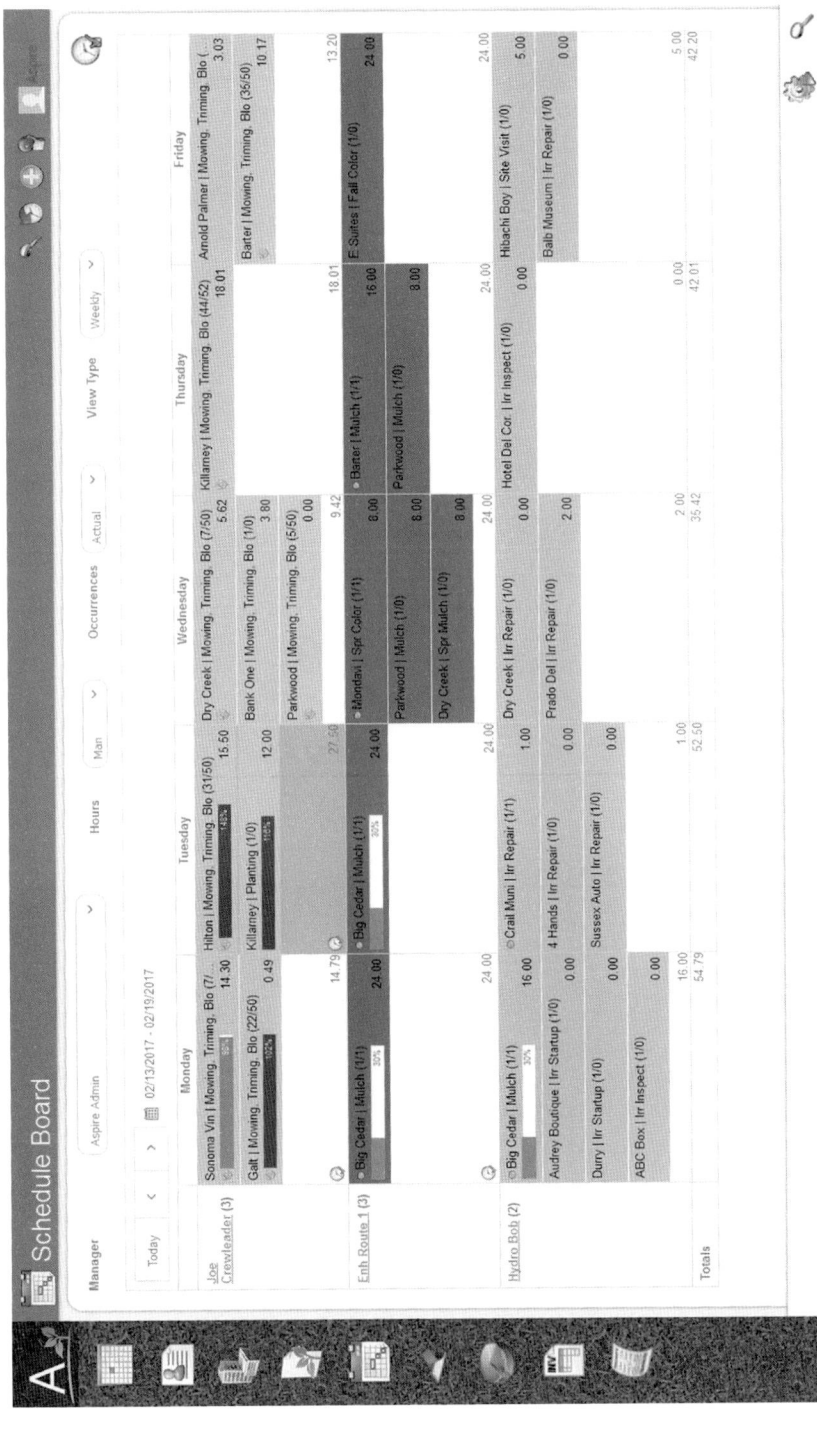

Figure 8.4 Schedule Board

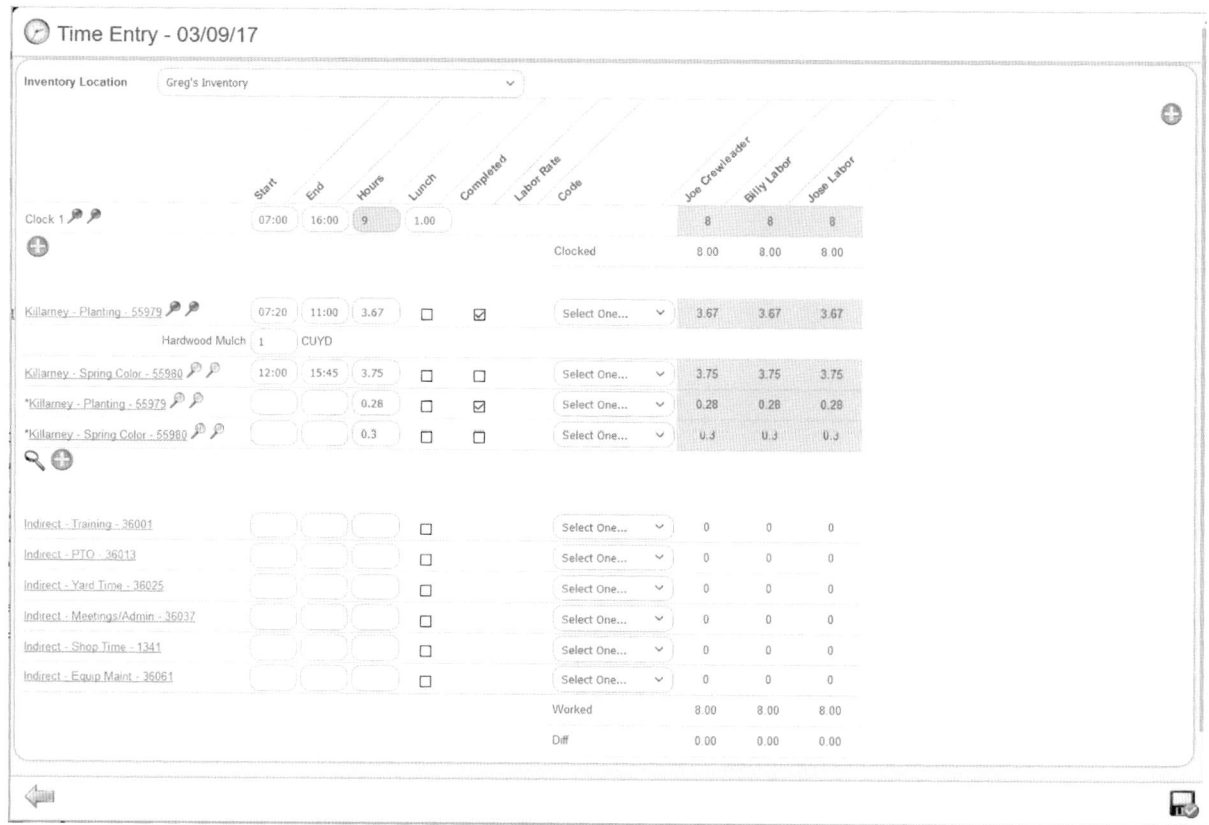

Figure 8.5 Time Sheet

Figure 8.6 Purchase Order

Report | Profit & Loss Monthly

PROFIT AND LOSS Monthly
Earned
January 2017 - February 2017

Groups		January 2017							February 2017							Totals						
Branch	Division	Revenue	Material	Equip	Sub	Other	Labor	Gross Profit	Revenue	Material	Equip	Sub	Other	Labor	Gross Profit	Revenue	Material	Equip	Sub	Other	Labor	Gross Profit
Main	Maintenance	$4,976	$100	$0	$0	$0	$1,484	$3,393	$1,589	$25	$0	$0	$0	$248	$1,316	$6,565	$125	$0	$0	$0	$1,731	$4,709
	Enhancements	$3,087	$750	$0	$0	$0	$454	$1,883	$1,259	$0	$0	$0	$0	$583	$676	$4,346	$750	$0	$0	$0	$1,036	$2,559
	Irrigation	$202	$0	$0	$0	$0	$75	$127	$542	$137	$0	$0	$0	$91	$314	$743	$137	$0	$0	$0	$166	$441
	Snow	$0	$0	$0	$0	$0	$0	$0	$0	$0	$0	$0	$0	$0	$0	$0	$0	$0	$0	$0	$0	$0
	Total	$8,265	$850	$0	$0	$0	$2,012	$5,403	$3,390	$162	$0	$0	$0	$922	$2,306	$11,655	$1,012	0	$0	$0	$2,934	$7,709

Figure 8.7 Monthly P&L

Software applications

| Report | Profit & Loss |

Aspire Landscape, Inc.
PROFIT AND LOSS
January 2016 - December 2016

Revenue	Maintenance	Enhancements	Irrigation	Snow	Total
Earned	$103,271	$171,886	$36,022	$80,112	$391,291
Credits	-$5,085	-$1,039	-$62	$0	-$6,185
	$98,186	$170,847	$35,961	$80,112	$385,105
Expense	Maintenance	Enhancements	Irrigation	Snow	Total
Labor	$14,153	$5,087	$2,078	$1,146	$22,464
Material	$93,265	$56,580	$7,748	$8,686	$166,280
Equipment	$0	$460	$0	$0	$460
Sub	$0	$475	$0	$1,039	$1,514
Other	$0	$0	$0	$0	$0
	$107,419	$62,601	$9,826	$10,872	$190,718
Gross Profit	-$9,232	$108,245	$26,135	$69,240	$194,387
Gross Margin	-9.40%	63.36%	72.68%	86.43%	50.48%

Figure 8.8 Annual P&L

Reporting systems

All accounting software interfaces with a reporting tool/system. Several programs are available for creating reports. Some of these programs have the capability of creating reports from any type of database and integrating them with Windows-based web applications. The reports can be formatted to print on preprinted forms or with customized fields and alignments.

Reporting programs format accounting information to assist management with day-to-day operations. They enable management to generate reports for any desired financial analysis, such as job cost breakdown by service lines (maintenance or installation), job cost analyses on a daily or weekly basis, monthly income statements, summaries of job proposals or estimates, or profit center budget projections versus actual cost comparisons.

Sales reports

Weekly update reports of sales booked versus sales goals are an important barometer for all sectors of the landscape industry. Sales reports are broken down into contract renewals, pending, and hot leads with probability, which are posted in weekly reports to provide current tracking and direction for the sales staff.

Each region of the country, based on its weather patterns, has a certain number of weeks during which landscape jobs can be scheduled. Therefore, regional and branch managers need to monitor the amount of business that is being contracted to stay on track for budgeted revenue goals. In the mid-Atlantic region, for example, jobs generally can be scheduled over a forty-two-week period. Maintenance companies in this region of the country strive to have 80–90% of their annual contracts by mid-April.

ESTIMATING SOFTWARE

Estimating software interfaces with databases of production standards, labor rates, material costs, overhead markups, and profit margins, for example, to generate job estimates.

Online take-off programs enable the user to trace electronic design files with their computer mouse and obtain square footage and materials quantities, which can be interfaced with estimating/bidding programs. These web-based systems expedite the submission of bids and job proposals. Maintenance estimating is facilitated with such programs as Go iLawn™ for determining square footage, plant and mulch quantities, equipment requirements (mower size), and fertilizer and chemical needs.

Figure 8.9 Go iLawn
Source: GoiLawn.com

Software applications

Once the estimate/bid is finalized, it can be formatted into a job cost budget that allocates direct costs over the duration of the contract. Field reports can be generated on a daily or weekly basis to provide supervisors with labor and material allocations.

Interfacing the estimating system with the reporting system enables tracking of the job cost status and its projected cost at completion.

Reports that further assist job cost management provide pertinent information on material purchasing (procurement), summaries of actual versus budgeted material costs, and documentation of project contract history, past transactions, and current status.

An additional feature offered by some project management programs is barcoding. Barcoding systems are used for work orders, enabling scanning into the production database without the need of manual data entry. Invoices and statements also can be barcoded and scanned directly into the client's file.

SMARTPHONE APPLICATIONS

Today's smartphones eliminate the need for managers to have field access to laptop computers. The plethora of applications available on smartphones enables them to maintain communication, track real-time job costs, transmit client information, access accounting files, download from the Internet, and even submit digital images of plants for disease diagnostics. Here is a brief summary of how applications can be implemented in a landscape contracting business:

- **Email communication**
    - clients – problem solving, proposal submissions
    - office – transmitting time sheets
    - vendors – product availability, delivery schedules
    - recruiting – responding to candidates, scheduling interviews, reviewing resumes
- **Photo and video camera**
    - client installation updates
    - plant and product selection proposals
    - plant disease diagnostics
    - site conditions
    - construction site damage
    - site documentation to validate contract compliance
    - design concept development

Software applications

- **Software applications**
  - generate spreadsheets, reports, and presentations
  - generate memos and invoices
  - payroll management
  - calendar and project scheduling
  - access company files
- **Internet access**
  - research and source products
  - access company's network
  - monitor weather projections
  - make travel reservations
  - access trade association websites

In essence, a smartphone serves as a minicomputer. Whether it is used primarily as a communication tool or expanded to interface with a company's network, it is an essential accessory for landscape contracting companies.

DESIGN SOFTWARE

There are numerous computerized landscape design programs, some of which are AutoCAD® based. As with other software programs, the company's application needs will determine program selection. The new generation of design programs is multidimensional and provides realistic perspectives of landscape features. Some of the features that have been integrated into these programs include:

- AutoCAD integration, with 2-D graphics for drafting with preset symbols, layers, and modes
- Property surveys, web-based plan searches, plan issuance records, links to quotations
- Drawing sheet output (hand-drawn style) with links to elevation and perspective views
- Automated material take-offs from finalized design, prepared to include grading plans, plant installations, hardscape materials, irrigation designs, night lighting
- Built-in database and spreadsheet capabilities for generating estimating reports with links to drawings to update reports and schedules
- High-end presentations with 3-D color renderings of terrain, shadow studies, and animated movies
- Team designing and editing feature

Software applications

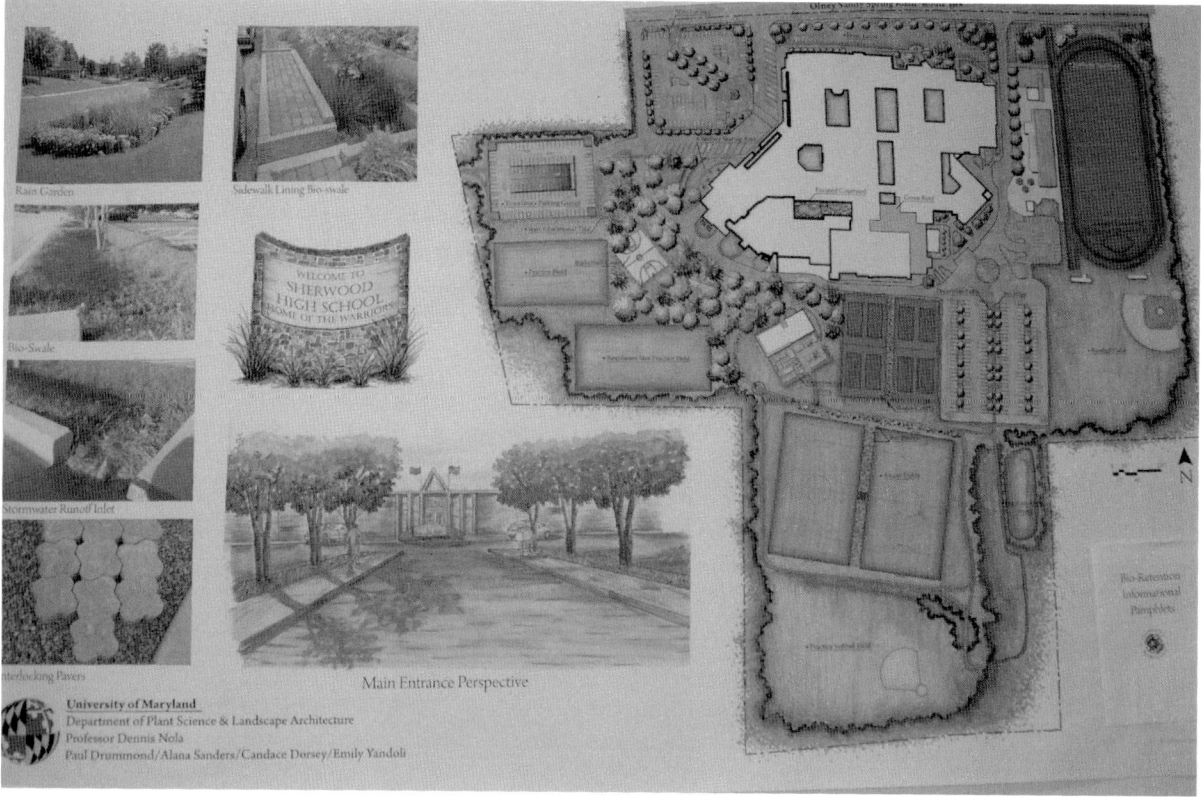

Figure 8.10 Digitized Landscape Plan Image

Design software enables a company to generate scaled multidimensional color renderings of its landscape concepts. These systems allow a footprint of the existing site – such as an entry drive – to be incorporated with desired features from the program's database, such as paver patterns and mature perspectives of plant materials. These databases may be expanded with the user's images of plant and hardscape materials. They also provide the flexibility of readily altering the renderings based upon changes requested by their clients.

MOBILE RESOURCE MANAGEMENT SYSTEMS

Systems that enable real-time tracking through the integration of GPS technology with business management systems are referred to as **mobile resource management (MRM)** systems. Most of these systems use wireless communication with the web through cell phones and text messaging, enabling transmission of data collected by GPS units. This web access

Software applications

enables branch offices to print out their reports through one centralized server. The MRM system provides managers with the opportunity to view a job in progress and assess its productivity based on budgeted labor hours. This information technology system is an excellent example of real-time management.

Routing and tracking

As landscape companies expand their geographic service area, tracking and routing become cumbersome. Through the use of innovative technology, this task has been transformed into an efficient and accurate computerized information system thanks to the U.S. government, which has made its **global positioning system (GPS)** available for civilian use. GPS was funded, developed, and operated by the U.S. military for defensive purposes. Through this system, coded satellite signals are transmitted and processed by GPS receivers that translate the signals into global position coordinates, velocity, and time.

The integration of GPS technology and mapping software has enabled companies to maximize crew efficiency in routing and scheduling clients. Downtime is reduced by utilizing the most direct routes, particularly during periods of construction, and by coordinating the scheduling of clients in close proximity to one another. Managers also can track how long a crew has been on a job site as well as the location of vehicles and their travel times to the site. Reports can be generated at the end of the day by downloading GPS units manually into PCs or automatically via receiver units. Reports generated from these programs document employee hours spent at specific job sites, travel time, travel speeds, and hours in service.

Routing software has the capability of reorganizing routes in minutes in the event of client cancellations or other factors arising, such as weather or hazardous situations. This information can be transmitted directly to supervisors in the field. Managers can actually transmit messages and scheduling updates instantaneously from their office PC. The program also offers the opportunity to generate reports, designating all jobs that are currently in progress within a specific radius. This feature enables routing in a logical sequence, thereby reducing travel time. This is another example of real-time management and maximizing production efficiency.

Mobile mapping

Another technological innovation for the landscaping industry has been developed for monitoring plant inventories on properties. The technology utilizes a positioning system that is mounted on the back of a pickup truck. It consists of a dual-frequency, dual-constellation

Software applications

global navigation satellite system (GNSS) receiver that establishes geospatial position of the vehicle; an **inertial measurement unit (IMU)** tracks the vehicle attitude (pose); a long-range **light detection and ranging (LiDAR)** sensor head equipped with 64 lasers that captures high-resolution, high-density point clouds at up to 100 meters away that are time stamped using special software; and external wheel encoders that capture odometer data from the vehicle. In addition, a high-resolution digital camera captures 360-degree images.

Figure 8.11 Mobile Imaging
Source: Topcon, topconpositioning.com

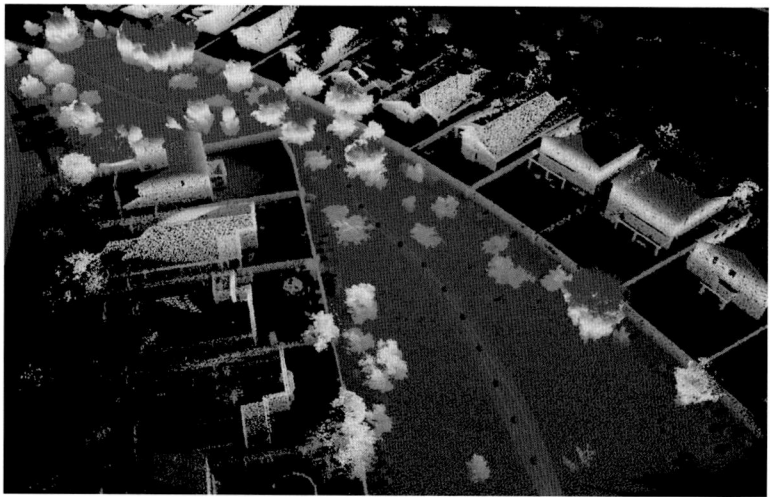

Figure 8.12 Plant Inventory Image
Source: Topcon, topconpositioning.com

Software applications

An actual application of this high-tech system was employed by a landscape company in Arizona. This company had already been using a geographical information system (GIS) to generate inventory databases for tree plantings at its maintenance properties, and the information was used to maintain the health status of the inventory and to schedule maintenance procedures. The company incorporated this information into a mobile mapping system that enabled the company to obtain a new perspective on its properties – plant density.

Plant density in an arid climate significantly affects water consumption. By using the GNSS system, the landscape company determined that a high-density property required 1.09 feet of water per acre, in contrast to a lower-density property that required only .19 feet of water per acre. The property with a higher plant density needed 355,179 gallons per acre, and that with the lower density 60,368 gallons per acre. At a cost of $4 per 1,000 gallons, the water cost would be $1,421 per acre for the high-density property versus $241 per acre for the lower-density property. Based on this information, modification of high-density properties can be proposed that will provide the company's clients with economic benefits as well as conserve a limited environmental resource.

In a climate that only receives eight inches of annual rainfall, it is vital that landscape companies efficiently manage the volume of water used in irrigation systems. The latter can be designed accordingly with information derived from a mobile mapping system. This is applicable to plantings on existing properties as well as landscape designs for new properties.

SUMMARY

Digital computer technology is an integral part of business management that automates data processing and provides instant visualization via reporting formats that can be shared between users. Virtually every aspect of business management can be accommodated efficiently and effectively with computer software and applications. Accounting systems are the most important of the various computing tools available, with mobile data processing and GPS tracking systems following in importance. Investing in computer technology benefits a company by enabling greatly improved operation efficiency and reduction of overhead costs, which positively affects the bottom line.

What are some software applications for the landscape industry?

- **Accounting** – monitor company profitability in branches and profit centers
- **Budget management** – track actual versus budgeted projections
- **Job cost management** – track job production costs in real time

Software applications

- **Estimating** – generate estimates, job budgets, field reports
- **Collection management** – track accounts receivable
- **Mobile resource management** – tracking, routing, and scheduling
- **Design** – plan development and client presentations
- **Mobile mapping** – plant density assessments
- **Payroll management** – real-time reporting

How are business management modules integrated through software programs?

Software programs are icons of production efficiency that expedite data entry and financial reporting and enable management to analyze and implement production efficiency measures. Generated reports provide real-time financial data that enables management to be proactive in budget management. Applications for smartphones also enhance the accuracy and timely processing of job cost data, scheduling updates, work orders, invoicing, estimating, and inventory management.

Tracking is a keystroke away with the development of real-time monitoring via GPS units, which incorporate software for data entry and transmission via the Internet. Managers can be at their office PCs while tracking jobs in progress and identifying areas for adjustments. Systems that incorporate real-time tracking are referred to as mobile resource management systems.

KNOWLEDGE APPLICATION

1. Based on your daily encounters with business entities, identify where information technology could be implemented to enhance operation efficiency.
2. Develop a new information technology concept that would be applicable to the landscape industry.
3. Describe how web access can be used to communicate with a landscape company's client base.
4. As the owner of a landscape maintenance company in your region, how would you utilize a mobile mapping system to serve your clients?

NOTE

1. Bruce Hunt, 2005, Former Vice President Brickman Group, Ltd. (currently BrightView), Personal Communication.

9

Managing human assets

CHAPTER OBJECTIVES

To gain an understanding of:

1. Establishing employee identity through job descriptions
2. Job description components
    a. Responsibilities
    b. Functions
    c. Qualifications
    d. Physical requirements
    e. Licensure/certification requirements
    f. Communication skills
    g. Specific job skills
    h. Education requirement
3. The recruiting process – media and venue options for posting job positions/candidate profiles
    a. Trade association websites
    b. H-2B recruiting agencies
    c. College career fairs
    d. Internships
4. Hiring legalities
    a. Comply with federal anti-discrimination laws
    b. Clearly define job requirements
    c. Enforce medical exam and drug testing policies for all candidates

Managing human assets

5. Creating a company culture
    a. Positive, caring environment
    b. Employee recognition
    c. Professional development
    d. Incentives
    e. Team engagement
6. Procedure for terminating employees
    a. Last resort
    b. Assessment of employee's evaluations
    c. Improvement recommendations
    d. Timeline for improvement status
    e. Clearly define reasons for termination
    f. Conduct an exit interview
    g. Obtain employee and witness signatures on a termination form

KEY TERMS

Americans with Disabilities Act (ADA)
internship programs
interview process

Landscape companies that have experienced growth and profitability can attest to the validity of viewing employees as company assets. Whether it is a design/build, construction, or maintenance business, management is cognizant of the fact that it isn't the equipment that returns dividends, it is the person operating the equipment.

The manner in which the human assets are managed is a determining factor in the growth and profitability of a company.

A great deal of the value of a company lies between the ears of its employees.[1]
   Front line is bottom line.[2]

Unlike capital assets, which can be depreciated and readily replaced, personnel assets must be appreciated and are not easily replaced, particularly in the landscape industry. Furthermore, when a human asset leaves a company, his or her value often is transferred to a competitor. Retention of these assets is contingent upon company culture, a topic that is discussed in the latter part of this chapter.

Managing human assets

An integral part of a company's culture is the recognition of employees as links in its chain to success. Identifying these links in job descriptions enables these individuals to know what their responsibilities are and what their role is within the company.

JOB DESCRIPTIONS

The job description provides an identity for employees and clearly defines their responsibilities and functions in their positions. It also provides employees and supervisors with indices of job performance and salaries. The description encompasses the qualifications required for the specific job title. Examples of qualifications for landscape industry positions would include the following:

- Physical abilities, such as lifting requirements
- Supervisory skills
- Equipment operation abilities
- License/certification requirements
- Multilingual communication skills, such as Spanish and English
- Sales experience
- Financial management experience
- Design skills
- Computer skills
- Estimating and bidding experience
- Bachelor's or associate's degree in horticulture or landscape contracting/management

The primary components of a job description are:

> 1. **Title** – specifically related to the responsibilities of the job
> 2. **Summary** – a concise summation of the job in terms of its functional role in the company
> 3. **Accountabilities** – a list of expected results for management and tasks for non-management personnel
> 4. **Physical requirements** – as they relate to compliance with the Americans with Disabilities Act (ADA) and other physical criteria
> 5. **Working conditions** – the physical working environment as it pertains to field and office personnel, such as multitasking, weather elements, noise, fast-paced, team-oriented, overtime, and weekends
> 6. **Supervisory responsibilities** – job titles responsible for supervising

Landscape Maintenance Company

Job Title: Production Manager
Department: Commercial Maintenance
Reports To: Division Manager
Preparation Date: March 25, 2011
Approved By: David Thomas
Approval Date: March 30, 2011

Summary

The primary purpose of this job is to oversee a group of crews and provide support and coordination for assigned crew supervisors. Ultimate responsibility: ensuring budgeted and quality performance of assigned projects.

Essential duties and responsibilities include the following. Other duties may be assigned.

1. Adhere to company policies.
2. Maintain a professional appearance and attitude.
3. Desire to increase industry knowledge and its application.
4. Desire to grow and advance with the company.
5. Perform required duties in an efficient and quality-oriented manner. Duties include various labor-oriented tasks in our various profit centers.
6. Review and grade subordinate employees on a monthly basis.
7. Recruit employees.
8. Be involved with marketing and sales.
9. Train assigned crew supervisors.
10. Deal with personnel issues.
11. Be involved in success and growth of the company as a top priority.
12. Assist in snow removal.

Supervisory responsibilities

Manages 1–5 subordinate supervisors who supervise a total of 6–20 employees in their landscape crews. Responsible for overall direction, coordination, and evaluation of unit. Carries out supervisory responsibilities in accordance with the company's policies and applicable laws. Responsibilities include interviewing, hiring, and training employees; planning, assigning, and directing work; appraising performance; rewarding and disciplining employees; addressing complaints and resolving problems.

Qualifications

To perform this job successfully, an individual must be able to perform each essential duty satisfactorily. The requirements listed below are representative of the knowledge, skill, and ability required for such performance. Reasonable accommodations may be made to enable individuals with disabilities to perform the essential functions.

Educational/experience

Bachelor degree from a four-year college or university; two years' experience or training in landscape management or a related field, or an equivalent combination of education and experience.

Language skills

Read, analyze, and interpret general business periodicals, professional journals, technical procedures, or governmental regulations.
    Write reports, business correspondence, and procedure manuals.
    Present information effectively and respond to questions from groups of managers, clients, and the general public.
    Command of conversational Spanish.

Mathematical skills

Calculate figures and amounts, e.g., discounts, proportions, percentages, area, circumference, and volume.
    Apply concepts of basic algebra and geometry.

Business skills

Interpret financial reports.
    Monitor and adjust financial components associated with profitability.

Reasoning ability

Define problems, collect data, establish facts, and draw valid conclusions.
    Interpret an extensive variety of technical instructions in mathematical or diagram form and deal with several abstract and concrete variables.
    Interpret financial reports and utilize the information for profit enhancement.

Certificates, licenses, registrations

Valid driver's license. Registration with the Department of Agriculture for application of pesticides.

Physical demands

The physical demands described are representative of those that must be met by an employee to successfully perform the essential functions of this job. Reasonable accommodations may be made to enable individuals with disabilities to perform the essential functions.

While performing the duties of this job, the employee is regularly required to work; use hands to finger, handle, or feel; and talk or hear. The employee frequently is required to stand, sit, and reach with hands and arms. The employee is occasionally required to climb or balance; stoop, kneel, crouch, crawl; and taste or smell. The employee must regularly lift/move up to 25 pounds, frequently lift/move up to 50 pounds, and occasionally lift/move more than 100 pounds. Specific vision abilities required by this job include close vision, depth perception, and ability to adjust focus.

Work environment

The work environment characteristics described here are representative of those an employee encounters while performing the essential functions of this job. Reasonable accommodations may be made to enable individuals with disabilities to perform the essential functions.

While performing the duties of this job, the employee is regularly exposed to outside weather conditions. The employee is frequently exposed to moving mechanical parts. The employee is occasionally exposed to fumes or airborne particles, toxic or caustic chemicals, risk of electrical shock, and vibration. The noise level in the work environment is often loud.

Physical demands refer to functions that are essential to the job and that cannot be reassigned to another position, eliminated, or accommodated with additional or artificial aid. The basis for this descriptive listing is to comply with regulations of the **Americans with Disabilities Act (ADA)**. These federal regulations specifically address physical, mental, and emotional job requirements. In the landscape industry, operating equipment and vehicles, lifting, and working outdoors would be listed under the physical requirements category; mental and emotional requirements would include working on multiple tasks, having good communication skills, and having the ability to cope with stressful situations. Nonessential functions are not listed to avoid being construed as discouraging disabled persons from qualifying for the position. For example, if a lawn mower operator is not required to lift heavy equipment, then lifting may be listed as a desirable, rather than as an essential, requirement.

Managing human assets

Job titles are standardized within the company to ensure uniform job responsibilities and designations in the company structure.

As a company expands through diversification or acquisitions, the titles may be revised. Since job titles identify an individual's role, careful consideration is given to the title designation. In some organizations, salespeople are referred to as account executives, crew members as team members, operators as technicians, and foremen as supervisors. Human resource managers recognize this as something other than semantics, since it enhances employees' self-image, which relates directly to their motivation and productivity.

Job descriptions have another essential function, which is to provide a profile of candidates for recruitment.

RECRUITMENT

"People are your most important asset" turns out to be wrong. People are not the most important asset. The right people are.[3]

Recruiting is a process by which companies attract the right persons for specific jobs. Recruitment emphasis in the landscape industry used to be primarily on management personnel until the advent of a labor shortage of field personnel. Based on a conservative 3.5% industry growth rate over the next twenty years, the industry projects that 600,000 new field employees will be needed by 2022. Newspapers no longer can be the primary resource for attracting them, and therefore a recruiting effort is necessary. Local employment agencies, religious institutions, high schools, and Hispanic media are contacted to recruit field labor. Financial incentives are often given to company field personnel to recruit employees. If the referral is hired and remains for a minimum period, the individual who recruited him or her is compensated. The recruitment of entry-level management personnel is equally as challenging. These candidates are usually two- and four-year graduates of landscape management/contracting programs. Currently and continuing into the future, there are not enough graduates to fill all of the entry-level management positions available. A concerted effort by landscape industry associations in which a positive image is projected for careers in landscaping would assist in attracting high school graduates to colleges and universities offering landscape contracting/management programs.

The U.S. government's H-2B program, until 9/11, had alleviated the shortage of field labor by providing a legal resource for migrant labor (see "Hiring" below). However, immigration quota restrictions and demands by other service industries have made it difficult for many landscape companies to meet their field personnel requirements. As Congress addresses immigration reform, there is hope that this situation will be improved.

Managing human assets

The recruiting process begins with management personnel who provide recruiters with detailed descriptions of the employment positions they have available and the job qualifications required. They also indicate the traits and skills necessary to successfully fill entry-level positions and those necessary to advance in the future. The recruiter then compiles the information into candidate profiles. A profile for a management position would include supervisory, financial, organizational, and communication skills, self-confidence, people skills, and an extroverted personality. The jobs are posted on company websites and circulated among universities, community colleges, and technical schools that offer curricula in horticulture, landscape contracting/management, and landscape architecture. There are also several green industry search firms whose websites may list job postings. When interviewing candidates, the recruiter attempts to identify key characteristics that will enable individuals to attain success in their positions. Also, they will assess an individual's attitude to determine if he or she will fit in with the company culture.

Recruiters seeking entry-level management candidates are present at university and community-college career days, industry trade shows, and the National Collegiate Landscape Competition (formerly Professional Landcare Network; www.landscapeprofessionals.org), a national event attended by more than 800 college students from all regions of the country. The career day venue provides a medium for initial screening of candidates' qualifications and personality traits. They are also observed competing in 28 landscape-related events ranging from equipment operation to business management and landscape design. Selected individuals are subsequently invited to interview at the company headquarters. The recruiter uses the same management-recruiting venues for hiring interns.

Figure 9.1 National Collegiate Landscape Competition (National Association of Landscape Professionals), 2013

Photo courtesy of author

Managing human assets

Figure 9.2 National Collegiate Landscape Competition (National Association of Landscape Professionals), 2013

Photo courtesy of author

Companies vying to recruit college graduates also host recruiting weekends. Student prospects are invited to spend a weekend in the vicinity of the company's regional or corporate offices. In this informal venue, the students have the opportunity to interface with management while becoming oriented to the company's culture and current career opportunities.

INTERNSHIP PROGRAMS

Internship programs are employed by the landscape industry to attract talented individuals who have the potential of becoming integral parts of the management infrastructure. The interns generally are employed for ten to twelve weeks during the summer. However, many companies offer internship programs year-round to provide interns with exposure to multi-seasonal operations.

Internship programs require a major commitment by management at all levels to provide a learning experience for the interns. The company designates an internship coordinator who is responsible for the program structure and intern assignments. This person is critical to the success of the program as a monitor and developer of rotations or training modules to meet the needs of the interns and to develop potential employees. This type of program also presents an opportunity to develop mentoring skills among the company's managers and supervisors.

Managing human assets

The following guidelines provide direction for implementing a successful internship program:

- Provide academic institutions with detailed information regarding the program's orientation and expectations
- Select a motivating coordinator
- Provide flexibility to meet interns' specific interests, e.g., design/build, seasonal enhancements, irrigation, maintenance
- Challenge the interns with projects that will enable them to apply, as well as develop, their skills
- Communicate with the interns on a frequent schedule
- Plan a company social event for the summer interns
- Provide a venue for obtaining intern feedback
- Select individuals as educational mentors
- Incorporate field trips to nurseries, botanical gardens, and high-profile landscape sites
- Expose the interns to local trade meetings
- Include interns in division meetings
- Provide competitive compensation
- Arrange or subsidize housing for out-of-state students

Figure 9.3 Student Interns
Photo courtesy of author

Some companies engage summer interns in landscape enhancement projects. Under the direction of a project coordinator, the interns schedule client meetings, conduct a site analysis, develop an estimate, present a proposal to the client in a sales presentation format, and complete the installation of the project.

Interns are an excellent resource for new talent. Companies have the opportunity to evaluate their individual talents as well as their potential as supervisors and managers. If the program is well structured, interns often return for subsequent internships, enabling them to apply their previous experience to entry-level supervision in addition to broadening their experience through exposure to other divisions of the company.

An example of a landscape industry internship program is a company that utilizes their client site at Busch Gardens in Tampa, Florida, to provide interns with a unique and horticulturally diverse learning experience. The interns are exposed to the latest developments in landscape management practices and have the opportunity to participate in numerous projects throughout the ten-week period. Areas in which the interns interface with staff include irrigation maintenance, seasonal color displays, turf management, pest management, sod installation, power equipment operation, and shrub maintenance, as well as general practices such as mulching, weeding, and fertilizing.

Chapel Valley Landscape Company in Woodbine, Maryland, determines the interests of first-year interns and provides them with a rotation through divisions that will enhance their skills. If students return for subsequent internships, they have the opportunity to fulfill their internship in a single division (for example, residential installation, maintenance, commercial installation, or irrigation).

Option I: Structured for the individual with limited hands-on experience.

- Two- to three-week rotation in commercial and residential installation and maintenance divisions.
- Emphasis on exposure to the industry through a variety of landscape projects.

Option II: Structured for the individual with at least six months landscape industry experience.

- Emphasis on knowledge and skill development.
- Individual selects one or two divisions for hands-on work experience.
- Additional experience is provided in business management through a one- to two-week rotation involving administrative projects. Individuals also have the opportunity to observe the role of management in daily decision making.

Managing human assets

HIRING

Legal concerns

A primary concern in the hiring process, in addition to selecting the right person for the job, is to avoid legal problems. Employer guidelines include:

- State specific desired skills and experience in job descriptions.
- Word ads carefully to avoid discrimination.
- Require written applications.
- Be careful with content and wording on application questions to avoid being charged for disability/gender/race/age discrimination.
- Don't make casual promises.
- Get applicant's written consent to obtain information.
- Check references and previous employers.

Mistakes in hiring can result in costly legal issues. Companies can avoid stepping into legal traps by describing jobs in compliance with Americans with Disabilities Act (ADA) regulations. As previously discussed, such compliance entails leaving out marginal or nonessential duties that may prevent a disabled person from applying for a position. Discrimination may become an issue if an ad is placed for a "salesman," which may be construed as excluding women. Reword the ad to ask for a "salesperson" or a "supervisor" rather than "foreman." Additional terminology that refers to age – "young" or even "energetic" – may be interpreted as discriminatory by the EEOC (Equal Employment Opportunity Commission).

Applications and interviews

Written applications are integral components of the hiring process. They provide a basis for documenting a candidate's qualifications. The applicant's signature on the application acknowledges that falsification of information will be grounds for dismissal. The signature also acknowledges, in an at-will employment state, a notification that employees can be dismissed without cause. References listed on the application should be authorized by the applicant, in writing, for permission to provide information. It is imperative that companies obtain additional information, such as credit checks, verification of academic credentials, driving records, police records, and previous job performance. Reference checks are important not only to verify performance but also from a safety standpoint. Drivers who are accident prone will be a liability, and a person who has a violent background may injure someone who could sue the company for negligent hiring.

The **interview process** can easily overstep legal boundaries, particularly since more than one interviewer is often involved. Interviewers should avoid sensitive areas such as marital status, children, health, age, race, or religion. Applicants may be questioned, however, about their ability to fulfill the job requirements and whether they would need special accommodations to accomplish it (for example, back support for clerical jobs). Medical exams for prospective employees are permitted by the ADA but only after a conditional offer of employment has been made. These exams must be required of all applicants, not just those who are disabled. Drug testing as a contingency for employment has become a common practice due to liability issues. This testing procedure must be enforced for all job applicants regardless of their position.

Foreign labor

Another legal issue affecting the landscape industry pertains to the employment of foreign labor. The landscape industry has been and will continue to be dependent upon foreign labor for field production. The federal government's H-2B program was initiated to assist companies. This program verifies nonimmigrants' legal employment status before they enter the U.S.

> The H-2B nonimmigrant program permits employers to hire foreign workers to enter the U.S. on a temporary basis and perform temporary nonagricultural services or labor on a one-time seasonal, peak load, or intermittent basis. Homeland Security regulations require that the petitioning employer initially apply for a temporary labor certification from the United States Secretary of Labor indicating that (1) there are not sufficient U.S. workers who are capable of performing the temporary services or labor at the time of filing of the petition for H-2B classification and the place where the foreign worker is to perform the work; (2) the employment of the foreign worker will not adversely affect the wages and working conditions of similarly employed U.S. workers.
>
> Section 101(a)(15)(H)(ii)(b) of the Immigration and Nationality Act (INA or the Act) defines an H-2B worker as a nonimmigrant admitted to the U.S. on a temporary basis to perform temporary non-agricultural labor or services for which "unemployed persons capable of performing such service or labor cannot be found in this country." 8 U.S.C. 1101(a)(15)(H)(ii)(b). Section 214(c)(1) of the INA requires DHS to consult with appropriate agencies before approving an H-2B visa petition. 8 U.S.C. 1184(c)(1). The regulations of the U.S. Citizenship and Immigration Services (USCIS), the agency within DHS which adjudicates requests for H-2B status,

require that an intending employer first apply for a temporary labor certification from the Secretary of Labor (the Secretary). That certification informs USCIS that U.S. workers capable of performing the services or labor are not available, and that the employment of the foreign worker(s) will not adversely affect the wages and working conditions of similarly employed U.S. workers.

Once the petition has been approved, the company applies for the work visas. Recruiting services (fee based) are often retained to bring employees into the country. The company is responsible for the employees' transportation costs and for locating a place of residence for them. Companies employing large numbers of H-2B employees often acquire or build housing adjacent to their businesses, which they rent to the employees. Many companies have benefited from this source of labor, particularly those who have been fortunate to have 80% of the previous year's H-2B employees return. The primary drawback to the H-2B program is the processing time. A 120-day period is required to file for employees. This requirement makes it difficult for landscape companies to anticipate their needs and to obtain employees when they are needed. Companies who apply for employees in November may not receive approval until March and then have to wait an additional thirty to sixty days for processing through the INS. However, this is the only program currently available, and in spite of the bureaucracy, it ensures that the employees are legally documented. Because the 66,000 quota for H-2B employees has not been increased since 9/11, the service-sector industry demand for these individuals far exceeds the amount that is available. Landscape companies are therefore becoming less dependent upon the H-2B program and relying more on recruiting from local communities.

The hiring process

A company's hiring process is standardized with specific procedures to ensure avoidance of legal issues and to consistently hire the best qualified candidates. The pool of candidates generated through the recruitment process is subsequently interviewed by management personnel and a human resources manager/director (recommended for companies with more than 100 employees). A human resources manager is responsible for administering personnel matters such as hiring, processing seasonal/immigrant labor, disciplinary matters, termination, benefits, professional development, and providing company manuals.

The interview process

The **interview process** is focused on gaining an understanding of the candidate in regard to personal goals, strengths and weaknesses, and experience that would enhance job performance.

Managing human assets

Companies often have testing agencies conduct profile testing. This type of test is thought to reveal psychological traits, such as the ability to deal with stress and challenging situations as well as other behavioral patterns. Management candidates often are requested to take a more comprehensive test to evaluate leadership, management, and personality traits. These tests are analyzed by the testing agency and a performance rating is generated.

1. Screening interview
2. Profile testing
3. Reference checks
4. Comprehensive interview

During the comprehensive interview, in which the candidate is asked specific questions pertaining to the position, a mutual decision may be reached regarding employment. The candidate is given a comprehensive understanding of the responsibilities associated with the position and his/her role as part of the company team.

Employee orientation

Following an employment commitment, human resources personnel give new employees an orientation to company benefits and company policies, which are enumerated in the employee manual. The manual is a communication tool that enables employees to understand the company's philosophy and policies, the obligations and expectations of employment, company methods and procedures, and employee benefits. It also addresses essential legal items such as EEOC guidelines, affirmative action policies, and COBRA (the Consolidated Omnibus Budget Reconciliation Act), which provides individuals and their families the right to continue their group health plan for limited periods of time after termination of employment. Coverage is predicated on voluntary or involuntary job loss, reduced hours, job transitions, and death, divorce, and sexual harassment complaint procedures.

EMPLOYEE EVALUATIONS

Employee evaluations are essential communication tools between supervisors and their employees. They enable the supervisor to evaluate the employee's performance of assigned job responsibilities and to make recommendations for improvement where deemed necessary. The employee has the opportunity to relate their respective needs in terms of training

Managing human assets

to enhance their performance as well as opportunities for advancement. A basic evaluation form (Figure 9.4) can be used for field personnel. Management employees are evaluated based upon the degree that they attained their annual goals. These evaluations should be scheduled on a regular basis, e.g., every six months. These evaluations serve as a record of performance and can therefore serve as a basis for advancement or as justification for termination.

EMPLOYEE EVALUATION FORM

Employee Name _____ Needs to Improve _____

Hire Date _____ Excellent _____

Date Last Increase _____ Good _____

Written Warnings _____ Fair _____

Attendance/Tardiness _____ Needs to Improve _____

Increase to _____ Probation _____

Excellent (E) Good (G) Fair (F) Needs to Improve (NI)

1. _____ Conforms to overall company philosophy and basic requirements for employment.
2. _____ Listens to and follows directions.
3. _____ Asks for clarification of job instructions.
4. _____ Requests demonstration or instructions on the safe operation of equipment.
5. _____ Communicates with fellow workers and supervisors.
6. _____ Eager to learn to new skills.
7. _____ Willing to work overtime and to assist with all tasks.
8. _____ Has a good working relationship with other employees.
9. _____ Self-motivated and capable of working independently.
10. _____ Completes assigned work assignments in a timely manner.

Comments

_____

_____

Employee Signature   Date   Supervisor Signature   Date

Figure 9.4 Employee Evaluation Form
Source: *Blueprint for Success*, National Association of Landscape Professionals, Herndon, VA

CREATING A POSITIVE COMPANY CULTURE

A company culture reflects its personality and includes its mission statement, values, goals, and overall work environment. An article by John Coleman, "Six Components of a Great Culture," in *The Harvard Business Review* (HBR Blog Network, May 6, 2013) lists the following elements:

- **Vision** – states the purpose of the company, which relates to the entire decision-making process within the company.
- **Values** – this gets to the core of the culture, providing guidelines on the behaviors and mindsets needed to achieve the company's vision. These values impact the manner in which the employees serve clients, interact with colleagues, and maintain professional standards.
- **Practices** – implementing the company's values through actions such as investing in employee development and continually providing them with care and respect.
- **People** – a company culture is dependent upon people who are willing to support and embrace the company's values. Recruitment should always be cognizant of candidates who are best suited to its culture. So-called "culture carriers" reinforce the organization's established culture.
- **Narrative** – a unique story that identifies, shapes, and retells a company's development and philosophy.
- **Place** – the work environment impacts the values and behavior of people in a workplace and shapes their culture, e.g., an open environment that incorporates atriums, outdoor landscape lounge areas, recreation rooms, and break rooms, etc., is conducive to employee interaction and collaboration.[4]

The company that adopts the philosophy that employees don't work *for* their supervisors but rather *with* them has established the basis for a positive company culture in which all employees are team members. The manifestation of its effects will be readily evident in employees' attitude and productivity. At all levels, employees want to know that they are appreciated and are playing an integral role in the company's success. Respect, positive reinforcement, recognition, empowerment, professional development, compensation, and company events are the key building blocks for constructing a business culture that will attract and retain dedicated and productive employees.

Employee retention

How long employees remain with a company, as well as their productivity, is directly related to the relationship with their supervisor/manager. Employees don't leave a company, they

Managing human assets

leave managers. Employees who leave take their talents and training with them, often to a competitor. In order to attract and retain talented employees, company management needs to address the issues that are of primary concern to these individuals. These concerns, enumerated in Chapter 1, "The Measuring Stick," of *First, Break All the Rules*, by Marcus Buckingham and Curt Coffman, include:

1. Do I know what is expected of me at work?*
2. Do I have the materials and equipment I need to do my work right?*
3. At work, do I have the opportunity to do what I do best every day?*
4. In the last seven days, have I received praise or recognition for doing good work?
5. Does my supervisor, or someone at work, care about me as a person?*
6. Is there someone at work who encourages my development?*
7. At work, do my opinions seem to count?*
8. Does the mission/purpose of the company make me feel my job is important?
9. Are my co-workers committed to doing quality work?
10. Do I have a best friend at work?
11. In the last six months, has someone at work talked to me about my progress?
12. This last year, have I had the opportunity at work to learn and grow?

*Most frequently cited among interviewed employees as links to retention.[5]

> *People don't change that much. Don't waste time trying to put in what was left out. Try to draw out what was left in; that is hard enough.*[6]

The manager's role is to motivate and develop each employee's talents, drawing out what is left in and thereby maximizing potential.

Successful managers have deviated from the norm regarding management style. A nonconventional or progressive manager approaches situations as follows, in contrast to a conventional style:

Conventional	*Progressive*
• Selects a person based on their experience, intelligence, and determination.	• Selects a person based on their talents and attitude.
• Sets expectations by defining the right steps.	• Sets expectations by defining outcomes.
• Motivates a person by helping him/her identify and overcome weaknesses.	• Motivates a person by focusing on his/her strengths.
• Develops the individual by helping him/her learn and be promoted.	• Develops the individual by finding the right fit, not necessarily the next rung on the ladder.

Talent is just potential; it takes managers to create an environment (culture) that allows talent to flourish. Progressive management defines the right outcome and lets each person find his or her own route to the outcome. An illustration of this concept would be a project manager who focuses on outcomes and avoids stopping and correcting each supervisor's management style. Defining outcomes encourages employees to take responsibility. This management approach places expectations on the employees while enhancing their self-awareness and self-reliance.

Successful recruitment and retention of young talent will depend upon a company's ability to challenge today's technically skilled millennials (gen-Yers), who were born between 1980 and 2000. These individuals are used to instant satisfaction and need to be constantly challenged. As entry-level managers, they should be mentored and given the opportunity to analyze operations and recommend productivity enhancements. Bill Cook (Figure 9.5) summarizes the priorities for meeting the needs of these individuals.

Managing millennials (gen-Yers) born 1980–2000

1. Provide structure – regular hours, timelines, agendas, clearly stated goals, regular assessment on progress.
2. Provide leadership and guidance. They want to learn and be mentored. They feel that they deserve your best investment of time in their success. They want to receive daily feedback and want to be in on the whole picture.
3. Encourage the millennials self-assuredness, can-do attitude and positive personal self-image. They are ready to take on the world. Encourage them . . . don't contain them.
4. Take advantage of the millennials comfort level with teams. Encourage them to join, they are used to working in teams. They have been raised to work in groups and play on teams. You can mentor, coach and train your millennials as a team.
5. Listen to the millennial employee. They are used to loving parents who have scheduled their lives around the activities and events of their children. These young adults have ideas and opinions and they don't want to have their ideas ignored.
6. Provide challenge and change. They are seeking ever-changing tasks within their work. What's happening next is of interest to them. Don't bore them, ignore them or trivialize their contribution.
7. Expect multitasking. They are multitaskers on a scale never seen before. They can talk on the phone while answering e-mails and instant messages. It's a way of life with them.

> 8 Take advantage of their electronic literacy.
> 9 Capitalize on the millennials' propensity for networking. Millennials like to network electronically beyond the work place.
> 10 Provide work-life balance. It's a concept new to this era. Millennials have been raised and educated in that philosophy. They are not into 60-hour weeks defined by the baby boomers. Balance in their work activities and their personal life is important to them and can make the difference in where they choose to work and how long they will stay.
>
> Source: *HR Insights*, Bill Cook, National Association of Landscape Professionals News, July 2009

Employee recognition

A basic human need is to feel appreciated. Recognizing this need and fulfilling it should be one of the primary objectives of a company culture. It can be easily achieved through positive reinforcement and recognition programs. Letters of commendation from the president of the company, employee-of-the-month recognition, and daily verbal recognition from supervisors instill a dedication and level of job performance that will exceed monetary compensation.

Recognition can take many forms, such as:

- **Tenure** – years of service recognized with pins or certificates, monetary awards, gifts at a company event
- **Birthdays** – cakes, personal greetings
- **Professional development training** – leadership/technical training workshops
- **Division competition** – recognition for attaining performance goals
- **Written recognition** – notes or letters expressing appreciation
- **Safety awards** – hours of service (crews) without accidents
- **Innovation awards** – implementation of production efficiency ideas
- **Client satisfaction** – acknowledgment from clients
- **Partners in excellence** – annual selective criteria
- **Informal** – daily acknowledgment
- **Formal** – acknowledgment in front of peers and companywide
- **Awards** – merchandise, gift certificates, dinners, trips, tickets to sports events

Recognition as listed above is often done at annual meetings, enabling all employees to recognize and honor their peers' accomplishments. This venue also serves as a motivating factor for individuals to strive to attain comparable recognition.

Managing human assets

A fun environment also adds to a positive working environment. For example, one advertising agency would surprise its employees with a monthly mystery bus trip. When the bus arrived, the office would be vacated and the employees whisked off to an amusement park, ball game, restaurant, and so on. The end product of a positive environment is a positive work force.

Another successful business example, The Walt Disney Corporation, has the following formula for business success:

Quality Cast Experience + Quality Guest Experience + Quality Business Practices = Success[7]
The primary components of Disney management style that contribute to its success are selection, training, communication, and care of employees.[8]

These components together comprise the Disney culture.

By celebrating success we create a culture of success.[9]

The Disney business culture is also comprised of standards in regard to job performance and client relations. Employees who feel part of a team and are motivated to perform their tasks will perpetuate these standards and the image that is created. The same principles hold true for a landscape company. The titles and job responsibilities differ, but the human elements are the same and so are the requirements: care, understanding, appreciation, and recognition as a team member. Care is a small word with a big meaning. All employees want to know that their employer cares about them as individuals and considers them an integral part of the company's success. The caring philosophy can be extended to the outdoor work environment: clean break rooms and restroom facilities, and a kitchenette for warming meals and storing food. Field personnel, who are often left out of the caring loop, will feel the positive impact of these efforts. Many landscape companies provide crew members with working attire for inclement weather and comfortable uniforms for warm weather, which is another significant caring attribute.

Jim Paluch, business coach and motivational speaker of his business management consulting firm, JP Horizons, feels that development of a team-oriented culture starts with:

- Identifying what the employee enjoys doing within the company structure
- Determining the projects and tasks at which the employee excels
- Focusing on training employees in areas that they excel at and enjoy
- Enabling employees to serve as mentors as a means of attaining self-respect

This philosophy is manifested within many professions, including sports. Former coach John Wooden of UCLA basketball fame once said, "Do not let what you cannot do interfere with what you can do." The implication of this statement is that management should reinforce the positive attributes or strengths of employees rather than dwell on their weaknesses. An illustration of this point was expressed by football coach Lou Holz: "Put the right person in the right job now." He uses the example of a player who runs the 40-yard dash in 4.9 seconds, who is slow for the tight end position but fast for the tackle or guard positions.

Training and development is an important component of a company's culture, since it provides resources with which individuals can rise and succeed. In every company, there are people with aspirations and talents that would enable them to be a greater asset if given the opportunity. In one landscape company, for example, a Hispanic employee was recognized by his supervisor for his perceptiveness and initiative. This individual was provided the opportunity to learn English and to be trained in his area of interest. Within a few short years, he became one of the company's top estimators and salespeople.

EMPLOYEE TERMINATION

Even in a tight labor market, employers terminate 15% of all new hires. The reasons are attributed to bad work ethics and attitudes and inadequate skills. The best approach to dealing with potential employee termination is to avoid it through counseling and monitoring with the goal of improving the employee's performance or behavior, or to find an alternative role that the employee can perform.

Federal and state governments now categorize more than 91% of workers into protected classes: minorities, women, people over 40, handicapped (inclusive of stress, obesity, high blood pressure, and perceived disabilities), pregnant, and adoptive parents. This broad protective umbrella provides a terminated employee with grounds for a lawsuit against the former employer.

If there is no alternative to termination, the manager should review the following guidelines:

- **Understand the situation** – analyze the basis for termination and check with human resources to verify justification. If downsizing, document the decision to do so and why a segment of personnel is slated for termination. Do not rehire an individual to fill a job category that will be abolished.
- **Employee manual** – does it have a policy procedure? Is there a warning process? In the event of an EEOC charge, the employee manual will be requested to verify if a policy has been violated. If an employee manual is not available, documentation will be accomplished through interviews and verification of past practices.

Examples of employee manual policy clauses

- **At-will employment** – States that the term of employment is not for a specific period of time and that it may be terminated at any time by the employee or company without reason, cause, or advance notice.
- **Equal employment opportunity** – Elucidates policy to provide equal employment opportunities regardless of race, gender, color, religion, national origin, age, physical or mental disability, medical status, marital status, veteran status, or sexual orientation.
- **Harassment** – States company's commitment to providing a work environment free of harassment associated with sexual orientation, gender, race, national origin, and age, physical or mental disabilities. Also includes a no-tolerance harassment policy by non-employees with whom the company has a professional relationship. Defines what is encompassed by harassment and disciplinary actions that will be imposed upon those that violate the policy.
- **Complaints** – procedures for filing and appeals.
- **Work hours** – daily work hours, breaks, and lunch period.
- **Overtime pay** – compliance with state and federal regulations. Calculations on which weekly overtime will be based.
- **Resignation or termination payment** – when checks will be issued, predicated on the notice provided by the employee in the case of resignations. Termination payments are specified in terms of mail processing period.
- **Employee benefits** – who is eligible, when they are eligible for medical and life insurance in terms of period of employment, employee contribution and other benefits. This includes workers' compensation insurance, medical insurance, life insurance, retirement programs, and tuition/course fee reimbursement.
- **Vacation, holiday, sick leave, maternity leave, family-care leave** – lists the paid holidays that the company observes, explains vacation and sick day accruals, vacation scheduling and approval procedure, and stipulates eligibility and use of sick days and family-care leave.
- **Conflicts of interest** – what constitutes a conflict of interest in regard to an employee's personal endeavors with those of the company?
- **Drug-free workplace** – defines substance abuse, disciplinary action, testing procedures, and consequences of repeated offenses.
- **Employment termination** – stipulates violations and situations that may result in termination. Explains severance pay in terms of its applicability to involuntary termination.
- **Safety** – company's commitment to maintaining a safe work environment. Provides safety guidelines and disciplinary action for infraction.
- **Union process** – were the contract provisions, grievance procedures, and appeals process followed?

- **EEOC considerations** – is there an existing affirmative action plan? How is the termination affected by the procedures associated with the plan? Is there documentation of all terminations over the last twelve to twenty-four months? Did the terminations include employees who were minorities, women, employees over 40, or disabled workers?
- **WARN (Worker Adjustment and Retraining Notification Act)** – if there are 100 or more employees and one-third of the workforce was laid off, was a 60-day notice given and were local community officials notified?
- **Retaliation** – have any of the terminated employees filed lawsuits, EEOC charges, or grievances, which could be construed as retaliatory action?
- **Patterns** – are there intentional or unintentional patterns to the terminations, such as those associated with health benefit coverage claims?
- **Consistency** – has the company been consistent in providing notices? Is there a consistent pattern of providing severance pay?
- **Documentation** – is there a chronological timetable of events leading up to the termination, such as noted on time sheets?

The termination process is done by the immediate superior, not by the human resources manager, who acts only if large numbers of employees are being terminated. Termination should be scheduled to take place between Monday and Thursday before noon, never on Friday at 4:30, after a business trip, or before a holiday. The process should be direct, respectful, and considerate. The reason for termination should be expressed clearly and concisely, in a respectful and fair manner. The employee should be given an opportunity to respond, but this process should not offer a venue for disputing the termination decision. The employee may be offered the use of company resources, such as a fax machine and telephone, for a period of time for a new job search.

Following the termination process, the employee should be sent to the human resources department for an exit interview (Figure 9.6) and further processing (Figure 9.7). The same applies for employees who have resigned. The exit interview enables companies to identify employment trends over a period of time and to establish a basis for management problems. The interviewer should be someone whom the employee views as impartial.

Examples of exit interview questions

- How do you feel about the company?
- What was your working relationship with your supervisor?
- Was there sufficient training for your position?
- Was your job performance evaluation accurate?
- Did you feel that you were fairly compensated for your skills and responsibilities?
- Were you given opportunities for professional development?

- Did your supervisor provide positive reinforcement?
- Rate the following with excellent, good, fair, or poor: Company benefits, salaries, working conditions, management and supervision, advancement opportunities.
- What company changes would you recommend to improve employee retention?

EXIT INTERVIEW

Name _____   Date of hire _____

Social Security number _____   Last day of work _____

Position _____   Department _____

Reasoning for leaving _____

Ratings	Excellent	Good	Fair	Poor
Overall Rating				
Company Benefits				
Salary				
Working conditions				
Manager/Supervisor				
Advancement or training opportunities				
Other				

Additional comments on items above _____

What specific circumstances led to this end of employment? _____

Would you consider, at some other time, working for our company again? Explain: _____

Would you recommend our company to others for employment? _____

Describe your working relationship with your supervisor? _____

What advice would you give our company to prevent termination in the future? _____

Figure 9.5 Exit Interviews Reflect the Working Environment from Employee's Perspective
Source: *Blueprint for Success*, National Association of Landscape Professionals, Herndon, VA

EXIT INTERVIEW

Employee comments _____

_____

_____

Interviewer comments _____

_____

Return of Company Property

Item	Yes	No
Uniform		
Equipment		
Keys		
Radio/Phone		
Manuals		

Benefits Information

Note: Medical, dental and life insurance benefits cease on the last day of the month of termination. You may elect to receive COBRA by contracting the office and completing the appropriate forms.

Payment in lieu of vacation will be made to the employee only if the individual is entitled to vacation and the employee resigns with proper notice, is laid off due to lack of work, or retires within the calendar year.

If you are enrolled in the 401-K plan you will need to contact the office regarding the distribution of funds.

_____   _____   _____   _____
Employee's Signature   *Date*   *Interviewer'sSignature*   *Date*

Figure 9.5 (Continued)

The final step in the termination process is the issuance of a letter that states the date of termination without expressing reasons for the decision. Other employees who are directly affected by the termination should be notified, such as immediate supervisors. The employee's file should contain detailed records on the basis for the termination, a copy of the exit interview, and statement of severance pay issued, if applicable. References may be given without legal concern if documentation can be provided for the respective statements.

Another legal concern with employee termination is a potential lawsuit that accuses the employer of preventing the individual from gaining future employment. A company can avoid this risk by the manner in which it disseminates information to reference inquiries. Information pertaining to confirmation of employment dates, title, salary, and whether the individual would be rehired, is safe ground from a legal standpoint. If the question is asked why the employee would not be rehired, a statement referring to company policy or job performance

TERMINATION FORM

Employee Information

Date of Termination _____ Social Security Number _____

Name _____

Address _____ City, State, Zip _____

- ❏ I chose to voluntarily terminate my employment with (Company Name), effective _____
- ❏ Employee has been terminated due to lay-off, effective _____
- ❏ Employee has been fired, effective _____ for _____

Employee statement _____
_____

_____   _____
Employee Signature   *Date*

Insurance Information

The law requires that under COBRA, (Company Name) offers employees continued health coverage under certain circumstances, including that the employee pay full monthly premiums in advance to (Company Name).

- ❏ I am not covered under the company's health plan.
- ❏ I am covered under the company's health plan, but do not wish to continue coverage.
- ❏ I am covered under the company's health plan, and would like COBRA coverage. (Additional form needed)

Company Property Information

- ❏ Employee received no company property.
- ❏ The property detailed on the attached sheets has been returned in good condition to the company, and the employee has no responsibility.

_____   _____
Received by   *Date*

- ❏ The property listed below and detailed on the attached sheet HAS NOT BEEN RETURNED TO OUR COMPANY. Employee understands that payment for company property, as agreed, will be deducted from employee's final paycheck. In the event there are insufficient funds for full payment, employee agrees to pay the company in full for any balance due.

Property not returned _____
_____

_____   _____
Employee Signature   *Date*

_____   _____
Company Signature   *Date*

Figure 9.6 Termination Forms Record Pertinent Information Associated with the Termination Process
Source: *Blueprint for Success*, National Association of Landscape Professionals, Herndon, VA

may be provided. If the future employer seeks more information, the company should request that the applicant sign a release authorizing further information to be given.

No law requires a company to employ someone who cannot do the job. Terminations are still a necessary part of doing business to ensure a company's future growth and development.

SUMMARY

This chapter has covered a plethora of topics pertaining to acquisition and management of employees. The single most important concept that should be gleaned from the text is that employees are individuals who require cultivation, appreciation, and development, regardless of their position. The companies who cultivate their employees reap the harvest of growth and profitability.

How is employee identity established?

Job descriptions establish the identity of employees as individual members of the company team. Teamwork among the employees at all levels, from field personnel to management, is essential to the growth and profitability of a company. The job title and summation of responsibilities establishes the functional role of the individual team member in the company.

What are the components of job descriptions?

The job description enumerates qualifications, accountabilities, and tasks associated with the position. Management positions also have expectations stipulated in the position description.

The essential functions of the positions pertaining to physical, mental, and emotional criteria are presented in compliance with the Americans with Disabilities Act requirements.

What is the recruiting process?

The recruiting process focuses on attracting qualified individuals to a company. It entails using several resources ranging from the federal H-2B program for temporary foreign workers to college career fairs, internship programs, and websites.

Managing human assets

What legalities must be considered in the hiring process?

Hiring is a process that must avoid many legal pitfalls. The wording in job postings and ads must be in compliance with ADA regulations and abide by EEOC guidelines. The same legal concerns must be addressed during the interview process, such as avoiding issues of health, marital status, age, race, and religion. Newly hired employees must be provided with the company handbook and made aware of company policies, including grounds for dismissal.

What comprises a company culture?

The most important asset to the recruitment process is a positive company culture, one that maintains a caring environment for each employee and provides him or her the mentoring and opportunity to reach their maximum potential. Positive reinforcement from management involves accentuating the positive rather than the negative, with the end result being increased productivity. Recognition programs that acknowledge achievement of personal goals through job performance and tenure are another manifestation of positive reinforcement in a company culture.

What is employee retention?

Employee retention refers to long-term employment of qualified employees. Companies that have been successful in retaining reliable, outstanding employees are those that have established a positive working environment. These companies regard their employees as assets and provide them the opportunity to attain their maximum potential.

What are the procedures for employment termination?

Employee termination is also a process that must proceed along legal guidelines. The first step in avoiding legal issues is not to set termination as a goal. Counseling and mentoring may be provided in an attempt to change an individual's behavior. It may also be possible to find an alternative position for the individual to better suit the person's talents and interests.

If termination is necessary, the following procedures should be followed:

- Chronological documentation pertaining to specific grounds for dismissal and counseling to allow for improvement

- Compliance with EEOC guidelines
- Reference to company manual's outline of grounds for dismissal
- Exit interview with the human resources department
- Termination letter

KNOWLEDGE APPLICATION

1. Write a job description, incorporating key components discussed in the chapter, for a position that you have held.
2. A potential applicant for a crew leader position enters your office with the assistance of crutches. What questions would you ask during the interview?
3. Describe a company culture that would influence your selection of an employer.
4. An employee was recently promoted to crew leader. Since his appointment, the efficiency of the crew has decreased, resulting in overruns on the budgeted time allocations. Would you terminate this individual? What factors would you take into consideration in reviewing this situation?

NOTES

1. Marcus Buckingham and Curt Coffman, 1999, *First Break All the Rules*. NY: Simon & Shuster.
2. Disney Keys to Success Seminar, 1998, Memphis, TN.
3. Jim Collins, 2010, *Good to Great*. NY: Harper Collins.
4. John Coleman, May 6, 2013, "Six Components of a Great Culture," *The Harvard Business Review* (HBR Blog Network).
5. Marcus Buckingham and Curt Coffman, 2000, "The Measuring Stick," *First, Break All the Rules*, Chapter 1, Gallup.
6. Ibid.
7. Disney Keys to Service Excellence, Disney Institute.
8. Ibid.
9. Ibid.

10
Productivity basics

CHAPTER OBJECTIVES

To gain an understanding of:

1 Production factors
    a   Inputs
        i   Capital
        ii Labor
        iii Equipment
        iv Materials
    b   Production management
        i   Production standards
        ii Scheduling
        iii Routing
        iv Skill development
        v   Team development
        vi Motivation
        vii Positive reinforcement
    c   Output
    d   Services
2 Production inefficiencies
    a   Travel time delays
    b   Equipment breakdowns
    c   Unskilled production personnel

Productivity basics

      d   Fueling
      e   Poor communication
      f   Not meeting job specifications
  3 Management's role in improving production efficiency
      a   Employee training
      b   Developing team environment
      c   Providing positive reinforcement
      d   Recognition of employee achievements
      e   Empowering employees
      f   Developing production standards
      g   Enhancing communication
  4 Process management
      a   Assessing process efficiency
      b   Defining inefficiencies
      c   Propose alternatives for improving the process efficiency
      d   Establish quality standards
      e   Train employees to meet the standards
  5 Integrating technology into landscape production systems
      a   GPS for routing . . . tablet and phone apps
      b   Estimating software applications . . . tablets, phones
      c   Job cost management systems . . . tablets, phone apps
      d   On-site invoicing systems

KEY TERMS

communication	PDCA cycle	productivity
incentive programs	process flowcharting	quality standards
kaizen	process management	

Productivity is a process that converts human and material resources into products or services. The efficiency with which this process is implemented determines a company's profitability. Payroll costs constitute the largest expense category on a landscape company's income statement. Attaining profitability, however, comes not from reducing the payroll expense but rather from increasing employee productivity.

Like the impact of robotics on mass production, the impact of mulch-blowing machines, leaf vacuums, and compact utility loaders have reduced muscle power and increased production

Productivity basics

efficiency in the landscape industry. The focus for achieving productivity efficiency is on working smarter rather than harder.

The landscape industry's emphasis on productivity is often directed toward field personnel, since they are performing the service that is sold to the client. In reality, productivity needs to be emphasized throughout the company since it affects every aspect of the company's operations. Management, administrative staff, sales, and estimating staff all have an impact on productivity. It is the brainpower throughout the company that enables the physical power to maximize production efficiency.

Figure 10.1   Toro equipment

Productivity basics

PRODUCTION FACTORS

In every job you have had, your performance has resulted in the production of a product or a service. Whether grilling hamburgers or mowing lawns, you knew what had to be done and what was required to get it done. You had a place to work and material resources to produce the end product or service – a cooked hamburger or a mowed lawn. In both job scenarios, a system existed with specific guidelines in place for you to complete the process.

The factors that contribute to a production process are illustrated in Figure 10.2. Production management provides the means for conversion of the input factors into a service. The efficiency of the entire process depends on what resources are available and how effectively they are managed to produce the service. The inputs are assets, which are the essential ingredients to implement the production process. Capital provides salaries and purchases the equipment and materials to provide the key components of the production process. These components are then incorporated into a production system that is structured to produce a service.

The competitive edge of a landscape company depends on the quality of its output and the amount of production hours required to produce it. The human element is the obvious key to achieving a company's production goal. Placing the right people in the managerial positions is critical to attaining production goals. These individuals are vital to the training and motivation of field production personnel. Depending on the job position, companies may use tests to analyze an individual's traits and determine strengths and weaknesses. There is a wide array of standardized tests that profile applicants and determine whether they would be a good fit for particular positions in management, sales, and so on. The tests match individual traits against profiles that are compiled after testing hundreds of people who are successful in their jobs. Some tests are also capable of determining an individual's energy level and compatibility. A sales profile would show a person who has a good self-image and an extroverted personality, is people oriented, and has strong communication skills. Management profiles consist of leadership traits, organization and communication skills, and most importantly, confidence in making decisions. What about field personnel? These individuals

Inputs →	Production Management →	Output
Capital	Production standards	Services
Labor	Scheduling	
Equipment	Routing	
Materials	Skill development	
	Team development	
	Motivation	
	Positive reinforcement	

Figure 10.2 Production Process

Productivity basics

are placed according to specific company profiles that include a desire to work outdoors and the ability to tolerate adverse conditions, the physical ability to operate equipment and perform tasks, the willingness to be a team player, and the receptiveness to learn new skills.

Management is responsible for the success of its production teams. It sets the parameters for the output and monitors the efficiency with which it is produced. Production standards for the company's operations are communicated and emphasized in the training of production personnel. The managers and supervisors are the team leaders who create the production environment, provide the resources, and set the goals for the output.

Table 10.1 Personality Test Illustration

Undesirable Behaviors	Test content	Desirable Behaviors
Slow to learn, needs simple tasks	**General mental ability**	Quick learner, clever, handles complexity well
Slow to figure out mechanical problems	**Mechanical reasoning**	Quick to figure out mechanical problems
Difficulty visualizing how parts fit together	**3-D reasoning**	Understands how 3-D objects fit together
Simplistic thoughts, negative attitudes, grammatical mistakes	**Open-ended questions**	Articulate, meaningful comments, expresses positive attitudes, no grammatical mistakes
Argumentative, critical, demeaning	**Agreeableness**	Pleasant, easy going, respectful
Uncomfortable in lead roles, easily bullied	**Assertive leadership**	Persuasive, influential, comfortable in lead roles
Fails to live up to promises	**Conscientiousness**	Lives up to obligations, adheres to rules
Overreacts to stress, gets upset easily	**Emotional resilience**	Calm, level-headed, handles stress well
Aloof, sparse communications	**Extroversion**	Keeps people well informed, builds relationships
Bends rules to suit own purposes	**Integrity**	Strong moral code
Direct, blunt, revealing	**Impression management**	Diplomatic, interpersonally sensitive
Doesn't feel responsible for how employees feel about their jobs	**Managerial human relations**	Good motivator for others, tries to help others succeed in their careers
Does not stay on top of tasks that have been assigned to others	**Managerial task structuring**	Stays on top of details that have been assigned to others
Sticks with tried-and-true approaches, resistant to new ideas	**Openness to new experience**	Enjoys learning and new ideas, innovative
Gives up easily, tends to find fault easily	**Optimism**	Positive about the future, good motivator, perseveres in spite of difficult challenges
Does not use time well, disorderly	**Orderliness**	Plans well, organized
Lacks confidence in own decisions	**Self confidence**	Confident of own decisions
Prefers working on own	**Teamwork**	Enjoys working in collaborative fashion
Focus on day-to-day problems	**Visionary leadership**	Focus on strategic, big-picture, long-term issues
Does the minimum	**Work drive**	Willing to go the extra mile

Productivity basics

PRODUCTION INEFFICIENCIES

Service businesses are time driven. The biggest challenge in providing those services is meeting the budgeted hours. Industry consultants have estimated that 35% of production time is wasted in landscape companies. Idle time or downtime (nonproductive time) can account for 15% of that time, most of which is wasted in the yard waiting for the crews to get their equipment and supplies for the day. Another example of time wasted is refueling vehicles, which idles an entire crew. Putting nonproductive time in perspective of lost revenue, a crew of three that is billed at a rate of $35.00 per hour and wastes 60 minutes per day will cost their company $7,000.00 annually in lost billable time (based on 40 weeks). Additional sources of lost production time occur with the following:

- Excessive travel time associated with poor routing
- Equipment breakdowns
- Return trips to the yard for materials, tools, and supplies
- Downtime while loading supplies
- Travel time to the job site
- Unskilled production personnel
- Lack of communication regarding budgeted hours
- Inability to communicate with foreign labor
- Insufficient briefing on installation specifications

PRODUCTION MANAGEMENT

A major portion of production management is associated with managing people. Managers need to be motivators and address the human factor rather than focus on the task-based factors. The fact that 19% of employees surveyed in a Gallup poll said that they felt disengaged at work reinforces this concept. These individuals work in a processing mode rather than focusing on results. The impact of disengagement on productivity is a loss of billions of dollars a year to employers. Motivating employees requires fulfilling their human needs and creating a working environment that will minimize their dissatisfaction.

Theories of motivation

Frederick Herzberg, in his research on employee motivation, developed a motivation theory that addressed minimizing employee dissatisfaction while simultaneously acknowledging employees and enabling them to advance within the work environment. The first part of his theory, called the hygiene theory, addresses factors that affect the work environment. All of the factors affect the satisfaction level of employees and their desire to achieve their maximum productivity potential. The extent to which management is cognizant of the employment impact of these factors determines whether a positive or negative work environment exists. These factors include:

Company policies and administration

- Policies communicated in a concise and comprehensible format
- Policies conveyed to employees with the goal of achieving understanding rather than intimidation

Supervision

- Emphasizing the positive rather than the negative, because positive thinking leads to positive results
- Providing opportunities for professional development; providing training to improve or acquire new skills; providing resources and training for certification exams, e.g., National Association of Landscape Professionals Landscape Industry Certified Technician and Landscape Industry Certified Manager, State pesticide applicator certification, and International Society of Arborists Certified Arborist

Working conditions and interpersonal relations

- Providing a clean and comfortable break room
- Providing the proper gear for adverse weather conditions
- Creating working relationships in which individuals feel that they are working with rather than for a supervisor
- Building a working team through trust and engagement in problem solving
- Instilling confidence and self-esteem by providing opportunities for self-accomplishment

Salary, status, and security

- Offering competitive salaries for comparable positions
- Offering compensation for advanced skills, experience, greater responsibility

Productivity basics

- Offering reassurance that the company is financially sound and employees' jobs are secure
- Providing employees with a job identity and a sense of playing an integral part in the company's success

The second part of Herzberg's theory involves creating a work environment that will foster the following motivating factors:

- Achievement
- Recognition of achievement
- Interest in the task
- Responsibility for enlarged tasks and challenges
- Opportunity for growth and advancement to higher level tasks

Management style is critical to implementing the factors listed above. Employing a participative system is the most effective in fulfilling human needs. The entire team plays a role in decision making, which results in striving to attain goals and raise the productivity bar. When every individual feels he or she has made a contribution, all employees' levels of self-esteem and self-confidence are elevated.

A 2003 Dartmouth College medical study concluded that all humans are "hard-wired to connect."[1] The study revealed that all of us have a need to communicate, to interact, to be motivated, and to motivate others. These results reemphasize the importance of addressing all employees' needs to motivate them toward a productivity goal.

MANAGEMENT SYSTEMS

Management systems are employed by companies to increase productivity, thereby improving profits and increasing customer satisfaction. Several articles and books have been written on the subject. Some are complex and require extensive training, others are more basic. The objective of any system, regardless of its complexity, is to get buy-in from all levels of personnel and to improve productivity efficiency. A brief synopsis of three systems follows.

Total quality management

Total Quality Management (TQM) is an integrative system that is structured to provide improvement within the business environment in response to internal and external needs.

The internal needs are in regard to improving company culture to meet the needs of employees and improve morale. TQM also addresses the human needs to attain job satisfaction, such as skill development and self-improvement. External needs refer to meeting customer needs, such as quality and responsive service. The needs are identified through daily management and integrated into a system of improvement. Long-term goals are set in the business plan to:

- Improve employee morale, skills, and productivity
- Improve quality of services
- Increase customer satisfaction
- Improve customer retention
- Reduce production costs

Six Sigma

Six Sigma is a management system that requires extensive training and the assistance of business consultants. The name represents levels of improvement, with Six Sigma being the highest achievable. The system is based on a problem-solving method that focuses on increasing customer satisfaction and improving profits by increasing production efficiency and reducing production costs. The foundation of Six Sigma is measurable results based on statistical measurements.

Six Sigma teams, comprised primarily of senior and middle management, are responsible for implementing process improvement. A process is defined as a sequence of steps associated with completing a task. The team members consist of the following:

Champion – a member of senior management whose responsibility lies with the logistical aspects and business principles of the process
Master black belt – who serves as a mentor to black belt team members, provides supports, reviews projects, and focuses on large projects
Black belt – the team leader, who has the primary responsibility of implementing the Six Sigma process
Green belt – an active participant in the project who engages in the implementation and application of the Six Sigma process

Process improvement refers to any business process, whether administration (e.g., invoicing) or field production (e.g., landscape installation). Every department is engaged in the improvement process. Priorities are given to those processes that have the greatest impact on customer satisfaction. Project criteria include:

- Clearly defined goals
- Project approved by management
- Manageable scope
- Project related to company mission

Team members pose the question, "How many mistakes do we make in the process?" After setting goals for the process – reducing downtime by 20%, for example – the team allocates resources to attain the desired goals. The team establishes the basis for the project by first justifying why it is a priority and deciding what its scope will be. It then proceeds to consider the following perspectives:

Financial – which objectives will ensure the financial success of the project?
Customer – what impact will this project have on meeting customer objectives?
Internal – what process needs to be addressed to achieve the project's goals?
Educational – what do team members have to learn and implement in order to achieve the project's goals?

Results are measured to track the improvement of the variations implemented in the process. This measurement also reaffirms changes that have resulted in improvement and thereby ensures that the same results will be attained in the future. The number of mistakes that occur within a process gauges improvement. The level of improvement ranges from a low of One Sigma, which represents a 30% perfection level, to Six Sigma, which represents a 99.9% perfection level. The Three Sigma to Four Sigma levels, which are within the 93% to 95% range, are where most companies generally plateau.

PDCA cycle

The **PDCA Cycle** (plan, do, check, and act) is a pretty straightforward process that mimics the previous management systems in a more simplified fashion. The cycle employs the following sequence of steps:

Plan – set a goal, e.g., reduce downtime by 20%.
Do – initiate an action for improving a process, e.g., fuel and load trucks at the end of the day.
Check – measure the results to assess the change in productivity efficiency, e.g., calculate the average crew downtime.

Productivity basics

Act – implement the new system through training and providing necessary resources. Determine whether additional action is required, e.g., designate individuals on a rotating basis to fuel and load trucks, or provide an additional pump on the premises to expedite the process.

As with all the management systems, the PDCA Cycle is an ongoing process, always setting new goals for improving the production processes.

Measurement indices serve as indicators of progress for everyone involved in the cycle.

People need to understand measurements in order to improve. You can't ask someone to lose weight without giving them a scale.[2]

Examples of indices are production hours for landscape operations compared to industry benchmarks, percentage of successful bids versus total submitted bids, percentage of contract sales versus sales calls, sales per full-time production employee compared to previous years, percentage of installation jobs completed on budget.

Incentive programs, which are tied to measurable goals, are catalysts for increasing productivity and profitability. These programs provide recognition and rewards for completing jobs efficiently. The success of incentive programs is contingent upon selecting goals that are within employees' control. Examples of such goals are finishing a job in fewer than the budgeted labor hours, improving safety records, exceeding sales goals, reducing customer response time, reducing downtime, and exceeding budgeted financial goals. They will be effective in stimulating production efficiency if they are tied to team performance as well as to individuals. Peer pressure can be an effective means of getting buy-in from all individuals.

Figure 10.3 Employee Recognition

Productivity basics

What types of incentives are effective catalysts? According to landscape professionals who gathered at a National Association of Landscape Professionals Executive Forum, they are not necessarily tied to monetary awards. Individual or team recognition on a weekly or monthly basis, gift certificates, extra time off, or attendance at an industry conference may suffice as a reward for performance.

IMPROVING PRODUCTIVITY

The best source for improving productivity is employees. They are engaged in their production roles on a daily basis, and therefore what better resource exists to determine how to improve a process? An individual (or a whole field crew) can be engaged in this exercise by outlining one of the processes and identifying the steps that could be improved to enhance production efficiency. This type of exercise is referred to as **process flowcharting**. The outcome of such an exercise is a detailed mapping of how the process is currently executed.

Involving employees in the process lets them use their initiative to discover how the process may be done more efficiently. Their knowledge mostly likely comes from their experience but may also reflect knowledge of what is being done at another company. For example, an accounting employee told her company about a software program that her husband used to track equipment maintenance in relation to lifetime hours. This program determined the efficacy of repairing a piece of equipment based on its current hours of productive service. Implementation of the program saved the accountant's company repair costs on equipment that was at the end of its productive life cycle.

Productivity efficiency should extend beyond field production. Overhead costs can be significantly reduced through implementation of more efficient administrative processes. Such administrative processes may include:

- Contract processing time
- Invoicing time
- Expediency of posting payroll and invoices into the job costing system
- Processing time for accounts payable and receivable systems
- Customer inquiry or complaint response time

Regarding the above processes, human or technical resources may be limiting production efficiency. Informal surveys of the landscape contracting industry indicate that companies have one administrative employee for every million to million and a half dollars of revenue. This ratio is not a hard and fast rule, since the number of personnel is relative to their level of productivity. However, companies whose administrative personnel ratios deviate from those mentioned should quantify their productivity and determine whether they are over- or understaffed. In this analysis, consideration should be also be given to the impact of technology, such as software programs on administrative production and to field technology; that would expedite data processing and thereby reduce the amount of data entry personnel needed.

Profitability is the name of the game and in order to achieve it companies must constantly assess the direct costs of delivering their service/product to their clients. Reducing labor costs, by increasing production efficiency, will increase gross margins, reduce overhead (associated with the reduction in labor), and thereby increase their net profit.

The discussion that follows addresses how companies can analyze and enhance their production processes.

PROCESS MANAGEMENT

Process management enables companies to compare the way a process was originally structured and how it is currently operating. These processes are documented and discussions are directed toward identifying steps that can improve their efficiency.

Production teams may be asked to address how to reduce production hours for mulching, for example. Teams who have the lowest production hours for this operation can illustrate through their step-by-step outline how this is accomplished. In an open forum, production teams can benefit from seeing how other teams overcome roadblocks that lead to inefficiencies.

This approach is similar to a process called **kaizen** used in Japanese industry. Kaizen encompasses incremental improvement through the power and knowledge of those who are involved in the process on a regular basis. This process evolved out of Toyota's Total Production System that focused on lean management discussed in Chapter 11 ("Creating a lean culture").

It relies on the knowledge and experience of these employees to identify where improvements are needed and to make recommendations for addressing them. Individuals on an installation crew may address crew size as an area in need of improvement. They might recommend that a crew size of three or four rather than six or eight would be more efficient, and further, that large jobs should have multiple crews, each with its own crew leader. In order for this process to succeed, each improvement needs to be:

- Measured to assess time and quality factors
- Monitored to assess efficiency and conformance to production standards
- Improved wherever deemed necessary

Productivity basics

Table 10.2 Productivity Standards Scorecard

Landscape Installation

Site: _____		Supervisor: _____	
Plant depth	____ (10)	Mulching	____ (10)
Rootball scoring	____ (10)	Watering	____ (10)
Bed edging	____ (10)	Pruning	____ (10)
Tree rings	____ (10)	Sodding	____ (10)
Staking	____ (10)	Cleanup	____ (10)
Deficiencies:			
Scored by: _____		Date: _____	

Landscape Maintenance

Site: _____		Supervisor: _____	
Mowing	____ (10)	Weeding (beds)	____ (10)
Turf color	____ (10)	Weed control (turf)	____ (10)
Trimming	____ (10)	Deadheading	____ (10)
Edging	____ (10)	Cleanup	____ (10)
Deficiencies:			
Scored by: _____		Date: _____	

Quality standards distinguish a company from its competition and are important in acquiring and retaining clients. Productivity improvement is, therefore, assessed on the basis of process efficiency within the respective quality standards. Jack Mattingly of Mattingly Consulting suggested developing a scorecard to periodically rate a production team. The scorecard breaks down the production process into its key components and rates each one on a point scale (Table 10.2). The parameters for each component are predefined to remove subjectivity from the rating.

Companies that have implemented process documentation have found that it is an effective means of identifying where roadblocks exist in a process and where breakdowns occur. When incorporated into the company culture, it serves as a reminder of what needs to be addressed to avoid inefficiencies that reduce productivity. All employees become proactive in seeking solutions for productivity improvement.

Training programs

Training programs are essential to productivity improvement. Employees who improve their skills or acquire new skills are capable of working more efficiently and productively. Their level of responsibility can be increased, and they become mentors for entry-level personnel.

Productivity basics

Training is also a motivator, since it advances the knowledge of employees and provides opportunities for their advancement.

Progressive companies are learning organizations. They are always seeking ways to become more efficient and productive. Some landscape companies require their employees to take a minimum number of education credits a year. The credits may be acquired through in-house training programs, community college courses, university extension short courses, or workshops and seminars presented by local or national trade associations, such as the National Association of Landscape Professionals, American Nursery and Landscape Association, International Society of Arborists, and American Society of Landscape Architects.

Certification programs provide training seminars to prepare candidates for the certification exams. The National Association of Landscape Professionals administers the Landscape Industry Certified Technician and Landscape Industry Certified Manager programs discussed in detail in Chapter 12. They require extensive preparation for application of skills and knowledge associated with interior and exterior landscape contracting. Some companies assign mentors to assist individuals in preparing for the exams. The end result of such programs is the elevation of professionalism in the field. These individuals become mentors for other production personnel and thereby establish an ongoing training program.

An effective mode of training is what is referred to as "just-in-time training." Information is presented to the employees just prior to when they can apply it. A pruning technique session on site prior to dormant pruning would be an example of just-in-time training.

Figure 10.4 Just-in-Time Training
Source: BrightView Landscape, Calabasas, CA

Productivity basics

Figure 10.5   Just-in-Time Training
Source: Davey, Kent, Ohio

Administrative costs account for a significant amount of a company's overhead and therefore warrant consideration as a training priority for enhancing production efficiency. Every administrative operation from payroll to contract processing can be addressed. An effective means of achieving administrative efficiency is through cross training. Cross training provides administrative personnel with new operational skills, enabling them to fulfill additional responsibilities. The benefits that will be derived include innovation by newly trained personnel who have an objective perspective of the new process and maintenance of productivity during the absence of key personnel due to illness, vacations, and attrition.

Cross training also has numerous applications to field production. Training installation and maintenance personnel in different production capacities enables them to work wherever they are needed for specific operations as well as to multitask on specific job sites. This additional skill training can serve as a motivator and a means of building self-esteem.

COMMUNICATION

Communication is also a vital entity to achieving production efficiency. Clear and concise communication between managers and supervisors ensures conveyance of expectations, current job status, and where improvements need to be implemented. Supervisors' communication with field personnel avoids misunderstandings and budget overruns from having to correct deficiencies in landscape operations. In addition, managers and supervisors affect productivity performance through daily positive reinforcement and recognition. The latter communication tools address the human factor, which relates to the need to be acknowledged and appreciated.

In today's landscape workforce, communication needs to be multilingual. Production processes must be instructed in the language of the crew members. Miscommunication often occurs when supervisors and managers are unable to convey instructions in the field. Productivity inefficiencies can be avoided through in-house training programs provided by college instructors or human resources personnel who are fluent in the languages spoken. The National Association of Landscape Professionals and other trade associations have published books with Spanish phraseology pertinent to the green industry. Colleges offer Spanish immersion courses, which emphasize communication skills and comprehension of Hispanic culture. The latter engages the human factor: understanding a culture in order to address individuals' needs.

In addition, communication of skills and an understanding of the whys and the how-tos will enhance productivity while elevating self-esteem. This pertains to conducting training sessions to convey the impact of maintenance and plant installation practices on the growth and development of plants, such as contrasting pruning with shearing and their effects on plant habits.

Productive communication is also achieved by maintaining a visual link with production schedules. These links are achieved by maintaining spreadsheets that reflect billable hours to actual hours, thereby revealing where inefficiencies occur and improvement is needed. Updating this data and providing copies to supervisors creates an awareness and impetus to improve productivity. Some companies post the information on job boards that reflect the budgeted hours to actual hours, such as mulching production hours listed by client with the actual hours expended. A combination of job boards and software apps will keep crews updated on their production status. Job boards are also used to post weekly and monthly job schedules. In the event of inclement weather, these postings enable adjustments to be made for labor and equipment allocations. The latter is particularly important for scheduling production sequences, such as fertilizing and mowing, grading and sodding, rototilling and seasonal color installation. Providing production crews with field job reports, budgeted hours, and materials and schedules provides communication links that will reduce downtime and improve production efficiency.

Communication is further enhanced with tablets and smart phones. This information technology is widely used by companies, providing the benefit of real-time communication with management and supervisory personnel. The latter contact is particularly important for reducing downtime (nonproductive), such as from equipment breakdowns. The nearest crew can transfer equipment. If the company has a mobile repair van, it too can be contacted for on-site servicing. Equally important is the system's ability to enhance client response time.

EQUIPMENT AND TECHNOLOGY

Equipment in the landscape contracting industry is the most important mechanical entity for achieving production efficiency. Breakdowns are the nemesis of production hours. Stringent maintenance programs and backup inventory reduce the downtime associated with breakdowns. If a company doesn't have in-house maintenance, then backup inventory is critical to avoid job delays due to slow turnaround from repair shops. Replacing heavily used equipment every three to four years affects production by providing the latest developments that will make job processes more efficient and thereby reduce labor hours.

The advent of zero turn radius mowers adds versatility to the mowing operation and reduces the number of hours associated with trimming. Mowers with chain rather than belt drives have substantially reduced required maintenance. Compact utility loaders like the Toro Dingo® (Figure 10.1) provide versatility in landscape construction. One machine that has the capability of trenching, digging ponds, lifting trees, transporting sod, and auguring holes for planting trees eliminates a substantial number of work hours as well as workers' compensation claims for strained backs. Power bed edgers enable crisp edges to be achieved without the need for shovel edging. They are also versatile enough to cut in tree rings. Contracting or purchasing mulch blowers greatly enhances productivity in large areas where wheelbarrows would normally be used to transport mulch to beds and tree rings.

Leaf vacuums not only increase productive efficiency over raking and bagging, but they also reduce the leaf debris to approximately one-tenth of its volume. The finely ground leaves can either be directly applied to beds or placed in compost piles to increase the nutrient value of the final decomposed product. Motorized fertilizer applicators increase the efficiency of application to large properties. They can be calibrated for both liquid and granular fertilizers.

Technology is being employed by the landscape industry to improve the production efficiency of operations in administration and field production. It has proven to be an asset for increasing accuracy and facilitating estimating, measuring, and accounting. Various design software programs have proven to be visual sales tools. Examples of specific applications are discussed here.

Client presentations

Digital imaging and 3-D design programs provide tools that enable clients to visualize landscape designs in perspective and on their residential or commercial footprint. User-friendly

Productivity basics

Figure 10.6 3D Digital Design

programs provide a database of plant materials and hardscape materials which can be utilized to generate a conceptual landscape design with an actual digitized image of the client's house or commercial complex. More sophisticated programs interface digital imaging with AutoCAD, providing a more realistic perspective of a landscape and in a multidimensional format, all of which can be presented to the client for a virtual perspective the landscape project.

Irrigation manufacturers provide CAD design services for their professional customers. Saving a company hours of design time and expense. These programs customize the design to meet the specifications of the client. They also incorporate water conservation with weather-based control systems.

Fuel consumption

Computerized fuel pumps provide accurate records of fuel costs for production vehicles and equipment. Coded cards are required to activate the system, which provides a database of the date and amount of fuel consumed. This information can be downloaded to the server to be incorporated into job costs for specific client accounts. In addition, this technology reduces

Productivity basics

downtime, since crews are not standing idle and spending nonproductive time in a gas station convenience store.

Sales and estimating

Information technology is enabling companies to manage sales leads, and enabling the sales team to assess their profitability and the potential of securing the contract. This assessment is based upon information that is posted with the leads regarding their status, e.g., new or existing client, and the scope of the project, e.g., softscape, hardscape, anticipated revenue, and the potential for obtaining the contract. The selected leads can then be interfaced with a bidding program that is interfaced with an on-screen takeoff system (utilizing PDF plans). A bid calendar is then generated and a schedule produced that has the respective job documents assigned to the individual bids. Since this integrated system is web based, the sales team and project managers can access the information at anytime and anywhere on their tablets or smart phones.

Maintenance companies can utilize software for field production management as previously discussed in Chapter 8.

Global positioning systems

The global positioning system (GPS), which was developed for the U.S. military, utilizes satellite signals to establish latitude and longitude. This system can detect ground locations within 300 feet. Since its deregulation in the 1980s, this location system has been adapted to a multitude of businesses. The landscape industry is one of the latest business sectors to implement this technology into production systems. Landscape companies equip their trucks with GPS units to add another efficiency quotient to their productivity formula. The system provides information on:

- **Arrival and departure time from job sites** – providing accurate information on time spent on the job and between jobs. It provides verification records to validate invoices and thus reduces the downtime associated with unscheduled stops.
- **Location of vehicles** – facilitating transferring equipment to other crews and directing mechanics needed to make on-site repairs.
- **Travel speed** – reducing safety risk, particularly with trailers.
- **Routing** – providing updated directional information to reduce travel time and avoid downtime due to construction sites and direction errors. The system can reroute crews in the event of emergencies, client requests, or cancellations. Interfacing with

a routing software program enables maximizing route densities and determining where deviations occur from the scheduled route. Routing efficiency reduces travel time and thereby increases productive time.

Linking the GPS unit with a networking feature called MRM (Mobile Resource Management®) expands the system to become a communication link with field personnel.

There are two types of GPS MRM systems, passive and active. The passive system collects the data and downloads the information at the end of the day. The unit can be removed for this process, or the data can be transmitted via an antenna mounted outside an office building. The active system provides constant communication links and transfer of information. This system enables wireless communication via e-mail between managers and crews, thereby providing personnel management. Last-minute client requests can be transmitted, as can any problems that are occurring on-site. Since the GPS is web based, managers can track the progress of their crews while remaining at their desk computer. If the crew is exceeding allocated time, the manager can determine what the problem is and convey a solution to the field supervisor. If it is a case of needing equipment or materials, the manager can arrange for securing whatever is necessary without involving downtime by having personnel leave the job site.

The productivity benefits of the GPS systems are:

- Accurate accounting of production labor hours
- Time management, since crews are cognizant of their production time
- Efficient means for managers to monitor productivity
- Reduced fuel costs due to routing efficiency

All of this technology requires a significant capital investment for the hardware, software, and monthly service charges (for active systems). However, based on productivity improvements and fuel savings, GPS system suppliers claim an ROI (return on investment) of over 300%.

Chemical systems

Plant growth regulators (PGRs) are chemical tools that are utilized by the landscape contracting industry to increase production efficiency. Their primary function is to reduce growth of turf, shrubs, and trees. Secondary functions include eliminating turf seed heads and undesirable fruit set on ornamental trees. When incorporated into landscape management programs, PGRs have a significant impact on the reduction of labor hours associated with mowing and pruning. The classes of plant growth regulators that have application in the landscape industry, as well as the effects of using PGRs, are listed in Table 10.3.

Productivity basics

Table 10.3 Plant Growth Regulators

Class	Growth Response	Landscape Application
Gibberellic acid inhibitors	Reduces stem elongation	Turf, shrubs, trees
Cell division inhibitors	Reduces plant growth	Turf, shrubs, trees
Ethylene enhancers	Prevents fruit formation	Shrubs, trees

Turf	Reduced cutting cycles, trimming, and edging
	Reduced thatch buildup
	Reduced water consumption
	Increased stress tolerance
	Restricted broadleaved weed growth (e.g., spurge, henbit, oxalis, clover)
	Transitions turf from annual rye to warm season grasses
	Suppresses *Poa annua* growth
	Restricts seed head formation
	Enhances turf seeding establishment through suppression of established turf
	Increases turf density among stolon and tiller turf types
	Improves color and appearance
Shrubs	Restricts shoots and terminal growth surges
	Enhances branching and shrub density
	Reduces or eliminates flowering and fruit formation (e.g., privet, multiflora rose)
Trees	Restricts terminal growth
	Reduces or eliminates undesirable fruiting (e.g., olive, crabapple, sweet gum)
	Removes mistletoe shoots
	Prevents seeding of mistletoe

The plant growth regulators alter plant growth through their effect on the plant's biochemistry. The classes of PGRs listed have two distinct modes of action, restricting the synthesis of plant hormones, such as gibberellins and cell division compounds, and enhancing the production of ethylene. Ethylene is often referred to as a ripening agent or maturation compound.

PGRs are most frequently used to manage turf on golf courses and large commercial sites. Lawnscapes of Ontario, California, evaluated the management impact of a growth regulator on a 165-acre site. After one year's application, the growth rate of the turf was reduced by 50%, resulting in a 26-week reduction in the mowing schedule, a 60% reduction in clippings and dump fees, reduced wear and tear on equipment, decreased fuel and equipment maintenance by approximately 50%, and a 20% reduction in water consumption. This company's use of a PGR enabled it to reallocate its crews from their previous 26 weeks of mowing to enhancing the appearance of other landscape entities.

Lawnscapes' net gain from PGR use:

Client satisfaction and retention – Reallocation of mowing time to landscape detailing, reduced water bills

- **Reduction of overhead** – lower equipment maintenance and fuel costs, fewer dump fees, and less downtime
- **Reduction of job costs** – fewer labor hours for clipping removal, trimming, and edging. A similar scenario occurs when growth regulators are applied to ornamental shrubs. One application of a growth regulator can save up to three months of trimming in warm season climates. This chemical tool reduces labor costs associated with trimming and removal of the clippings.

Plant growth regulators are a production management tool. It is not a tool that will necessarily have application to every job, and where it is applicable, the justification should be based on an economic assessment. Management needs to determine whether the material and application costs provide sufficient savings in labor and equipment costs to incorporate PGRs into a client's contract. One of the key cost determining factors is the length of the growth control period and the company savings during this period.

The basics of productivity are people related. How people are managed, trained, and motivated determines the amount and quality of their production. Providing employees with the latest developments in equipment and technology elevates the level of production that can be achieved. The key to productivity is management of a company's human assets.

SUMMARY

What are production factors?

The production process consists of three primary factors: input, production management, and outputs. The most important input component is labor, based on the impact on the quantity and quality of output. Production management is the manner in which the labor component is managed and provided with production resources. Production management is the catalyst for production efficiency.

What are some of the primary production inefficiencies?

The most consistent production inefficiency associated with the landscape industry is downtime or nonproductive time. This inefficiency is associated with delays in crews leaving the company yard, fueling, and routing. Additional inefficiencies are associated with equipment breakdowns and inadequate training of field personnel.

Productivity basics

What is management's role in improving production efficiency?

Management is directly responsible for identifying areas of production inefficiency and determining what is necessary to correct the situation. Rerouting or reassigning job sites could reduce travel time; reassignment of mower sizes used on job sites may increase efficiency; mechanizing tasks, e.g., with utility loaders, also may increase efficiency. Most importantly, management is responsible for motivating and training production personnel through positive reinforcement, recognition, and professional development.

What is the employee's role in increasing production efficiency?

Employees are the gears that turn the production wheels. Their role is to perform their tasks in the most expedient manner while maintaining company standards. Each of them has the responsibility of identifying inefficiencies and becoming a part of the solution. The solution may require training and development and/or additional resources, such as new equipment. Their engagement in process management processes is important to achieving the goal of production efficiency.

How can resources be maximized to increase productivity?

Resources can be maximized to increase productivity by management's analysis of the inefficiencies and identification of the resources required to correct the inefficiencies. Engaging personnel in analyzing landscape operations through a system such as the PDCA Cycle will aid in the identification of inefficiencies and the implementation of corrective measures to increase productivity. Corrective measures that address time management, development, motivation of human resources, and provision of required material resources will produce the output desired: increased productivity.

How has technology been integrated into landscape production systems?

Computer technology has had the most significant impact on the landscape industry. Real-time job cost reporting and estimating systems via software programs have increased efficiency and productivity. Additional applications include production of landscape designs and client presentations. The use of GPS units has further enhanced productivity through tracking, routing, and job cost management. Plant growth regulators are chemical tools that increase landscape management production efficiency. This class of chemicals includes growth retardants, pruning agents, and flower and fruit inhibitors.

KNOWLEDGE APPLICATION

1. List four distinct production-related deficiencies that you have observed among commercial companies or municipal agencies in your community.
2. Apply the PDCA Cycle to a landscape maintenance or installation operation.
3. If you were the owner of a company, how would you develop a culture that would motivate your employees to maximize their productivity?
4. Outline an incentive program for increasing production efficiency within a landscape construction company.

NOTES

1. Anonymous.
2. National Association of Landscape Professionals, 2002, *Crystal Ball Report #23*, Herndon, VA.

11

Creating a lean culture

CHAPTER OBJECTIVES

To gain an understanding of:

1. Lean management
   a. The 4 P's
      i. Philosophy: Long-term commitment to customer value
      ii. Process: Eliminate waste
      iii. People and partners: Work together to achieve lean processes
      iv. Problem solving: Team engagement
   b. Process
      i. 5 S's
         - Sort
         - Straighten
         - Shine
         - Standardize
         - Sustain
      ii. Develop a lean culture and reduce non-value-added time
      iii. Confront problems and come up with solutions
2. Kaizen events
   a. Identify and document process problems
   b. Propose and test alternatives
   c. Assess and establish new process standards
   d. Celebrate

Creating a lean culture

KEY TERMS

4 P's – philosophy, process, people and partners, problem solving

5-S process – sort, straighten, shine, standardize, sustain

kaizen

muda

non-value-added time

ratio of profit per dollar of sales

ratio of sales per dollar of salaries

Toyota Production System (TPS)

value-added time

Lean is a journey toward the constant pursuit of perfection.[1]

Developing a lean system is similar to investing for retirement. Effort and sacrifice needs to be in the short term to reap benefits in the long term.[2]

THE TOYOTA PRODUCTION SYSTEM (TPS) AND LEAN MANAGEMENT

The **Toyota Production System (TPS)** was developed by Toyota engineer Taiichi Ohno and Eiji Toyoda between 1948 and 1975 to improve Toyota's assembly-line manufacturing process as a means of increasing value to customers. The primary goal of TPS is to eliminate waste, called **non-value-added time**, which is manufacturing time that does not contribute to the end product. Non-value-added time existed in Toyota's assembly logistics, workers' physical access to parts, quality control, and replacement of defective parts, equipment downtime, and disruptions in the production flow. Ohno's re-engineering of the process achieved the following goals:

- Getting the right things to the right place at the right time, the first time
- Minimizing waste and being receptive to change

With this system in place, employees' time was converted to **value-added time**, time that contributes to the end product. The system eliminated waste, enhanced production flow, and improved quality. The ideology behind TPS became today's lean manufacturing.

How has lean management affected Toyota? The answer lies in its dominance as one of the leading car manufacturers in the world. While other car manufacturers have closed production plants, Toyota has expanded. Is the company's success associated with lean management? The answer is a resounding yes, and the culture that perpetuates its success is constantly striving to improve by reducing waste and increasing product quality.

It is the success of the lean manufacturing concepts behind TPS that led to its adaptation to the green industry. The National Association of Landscape Professionals (formerly National

Creating a lean culture

Association of Landscape Professionals) Crystal Ball Report #26 addressed this specific subject and is entitled "Lean Management for the Green Industry: An Operational Strategy That Delivers Value to Customers and Eliminates Waste." The committee that produced the report defined lean management in the green industry as "An operational strategy designed to radically improve customer satisfaction, labor productivity, and morale, quality, on-time delivery of service, and profitability, while rigorously eliminating waste and increasing value to the customer."[3]

Lean management is applicable to every process that occurs within a business operation, whether it is administrative or operational.

Administrative Processes:

- Payroll
- Accounts receivable
- Accounts payable
- Recruiting
- Hiring
- Performance evaluations
- Budgeting
- Estimating
- Job cost management
- Inventory management
- Equipment maintenance
- Purchasing
- Training
- Proposal development
- Contract management
- Subcontractor management
- Training
- Sales/prospect management
- Routing
- Scheduling

Operational Processes:

- Production standards
  - Mowing
  - Mulch rings
  - Mulching beds
  - Pruning
  - Planting – trees, shrubs, groundcover, annuals, perennials
  - Tree staking
  - Pesticide application
  - Watering
  - Fertilizing
  - Irrigation startup and shutdown
  - Hardscape installations
  - Snow removal
- Crew startup
- Truck loading/unloading
- Trailer loading/unloading
- Vehicle/equipment fueling
- Daily/weekly site scheduling
- Green waste management shift composting and recycling under green waste management as subheads
  - Composting
  - Recycling

When you think of anything we do in our daily lives, from the time we arise to the time we go to bed, there isn't anything in terms of a process that can't be addressed to maximize efficiency and reduce waste. Within our personal lives it most often relates to time management. Time reduction in the workplace relates to dollars saved and increased profit. Here is just one example: Terp Landscape Company has 70 crew trucks that leave the yard on a daily basis. Each truck has three crew members with an average hourly wage of $10.50 per hour (a total of $31.50 per hour). If the company were to reduce startup time (departure from the yard) by ten minutes (.16 hr) for each crew, it would have savings of:

Daily:	$352.80	($31.50 × .16 × 70)
Weekly:	$1,764	(5-day week)
Monthly:	$7,056	
Yearly:	$70,560	(10 months)

This simple example illustrates what lean management can accomplish. We'll see more complex applications later in the chapter after the lean management processes are discussed.

COMMITMENTS TO A LEAN CULTURE

Before a company embarks upon implementing lean management, executive management needs to make specific commitments. The Toyota Production System proposes the 4P Model, which are (1) philosophy, (2) process, (3) people and partners, and (4) problem solving.

Philosophy

The A company's long-term commitment to the lean culture philosophy – to function as a complete entity in providing value to the customer – sets the foundation for all the other principles. This commitment encompasses constant reassessment of the company's economic performance and growth.

Process

A commitment to process relates to eliminating waste by incorporating lean production methods throughout the company. Examples of waste in the workplace (Ariens Company, Brillion, WI) are listed in Figure 11.1.

It also places emphasis on striving for a value-added flow within its operations, not only to reduce waste but also to enhance the quality of the end product/service to the customer. Toyota management came to the conclusion that the right process will produce the right results. This is not limited to individual processes but also includes the impact of the right process on associated business entities, such as those that provide materials for production or deliver the end product or service. In a landscape installation scenario, what impact would the soil preparation process have on a landscape company's sod installation

EIGHT CLASSES OF WASTE

1. Waste of inventory:

2. Waste of transportation:

3. Waste of processing:

4. Waste of waiting:

5. Waste of motion:

6. Waste of over-production:

7. Waste of a defect:

8. Waste of a person:

Figure 11.1 Eight Classes of Waste
Adapted from James P. Martin, 2006, "Lean Management for the Green Industry," Crystal Ball Report #26, National Association of Landscape Professionals Herndon, VA

Creating a lean culture

if rototilling had only loosened the soil to a depth of two inches, with one-quarter inch of organic matter and superficial grading? Another example of a situation with process problems would be one in which seasonal color plantings are installed but the watering crew doesn't show up for three days.

People and partners

People and partners are integral components of the lean culture. People, in this case, refers to the company's employees, who must adopt the lean management philosophy to assure management that lean methods are implemented and maintained by all employees' respective processes/operations. Lean is not a process that is based on top-down decisions; its success depends on engaging employees at all levels in the lean decision-making process and developing their skills to enhance their contribution to working lean.

Partners refers to the suppliers whose products and services, e.g., delivery, have a major impact on the flow of service operations. An example is a nursery business that delivers seasonal color to a job site's enhancement crew. If the delivery is delayed or the quality of the plants is unacceptable, the enhancement crew accumulates significant non-value-added time. Partners are also important to the landscape industry when it comes to equipment. Manufacturers will often customize equipment to accommodate client needs or to respond to operational issues. An example would be the addition of a shield on cyclone spreaders for sites that are adjacent to waterways, as a means of reducing nutrient runoff. Lawn mower manufacturers have also responded to the need for maximizing efficiency by producing chain-driven instead of belt-driven mowers to reduce equipment breakdowns and by engineering more durable decks.

Problem solving

A commitment to problem solving is directly tied in with a company developing and engaging its people. Problem-solving efforts have to permeate the entire organization, from top to bottom. In many instances, it will be from the bottom up, as when employees alert management to problems with their operations. The key to success with this commitment lies in recognizing a problem and making a decision to rectify it. By engaging all employees in the commitment to problem solving, everyone gains from the learning experience. This engagement begins with the employees' prioritizing projects based upon a quantitative assessment of the criteria listed in Figure 11.2.

Creating a lean culture

PROJECT CALCULATOR

Owner: _____ Date: _____

Project	Importance to Customer	Cost to Implement	Feasibility Likelihood of Success	Cost Reduction	Leverage Positive Impact on Others	Quality Positive Impact on Quality	Delivery Positive Impact on Delivery	Total Project Priority
	Rate 1 to 5 – Circle One (High = 5, Low = 1)							
	1 2 3 4 5	1 2 3 4 5	1 2 3 4 5	1 2 3 4 5	1 2 3 4 5	1 2 3 4 5	1 2 3 4 5	1 2 3 4 5
	1 2 3 4 5	1 2 3 4 5	1 2 3 4 5	1 2 3 4 5	1 2 3 4 5	1 2 3 4 5	1 2 3 4 5	1 2 3 4 5
	1 2 3 4 5	1 2 3 4 5	1 2 3 4 5	1 2 3 4 5	1 2 3 4 5	1 2 3 4 5	1 2 3 4 5	1 2 3 4 5
	1 2 3 4 5	1 2 3 4 5	1 2 3 4 5	1 2 3 4 5	1 2 3 4 5	1 2 3 4 5	1 2 3 4 5	1 2 3 4 5
	1 2 3 4 5	1 2 3 4 5	1 2 3 4 5	1 2 3 4 5	1 2 3 4 5	1 2 3 4 5	1 2 3 4 5	1 2 3 4 5
	1 2 3 4 5	1 2 3 4 5	1 2 3 4 5	1 2 3 4 5	1 2 3 4 5	1 2 3 4 5	1 2 3 4 5	1 2 3 4 5
	1 2 3 4 5	1 2 3 4 5	1 2 3 4 5	1 2 3 4 5	1 2 3 4 5	1 2 3 4 5	1 2 3 4 5	1 2 3 4 5
	1 2 3 4 5	1 2 3 4 5	1 2 3 4 5	1 2 3 4 5	1 2 3 4 5	1 2 3 4 5	1 2 3 4 5	1 2 3 4 5
	1 2 3 4 5	1 2 3 4 5	1 2 3 4 5	1 2 3 4 5	1 2 3 4 5	1 2 3 4 5	1 2 3 4 5	1 2 3 4 5
	1 2 3 4 5	1 2 3 4 5	1 2 3 4 5	1 2 3 4 5	1 2 3 4 5	1 2 3 4 5	1 2 3 4 5	1 2 3 4 5
	1 2 3 4 5	1 2 3 4 5	1 2 3 4 5	1 2 3 4 5	1 2 3 4 5	1 2 3 4 5	1 2 3 4 5	1 2 3 4 5

Rate prospective projects in seven categories, and use the combined score for an overall ranking. Use estimates to set initial priorities and recalculate at a later date when accurate data is available.

Figure 11.2   Project Calculator
Source: Ariens Company, Brillion, WI

STARTING WITH THE 5 S'S

The 5 S's refers to a process that in essence gets our house in order. Whether we are addressing our desks, the maintenance garage, trailers, or trucks, the principle is the same. Unless there is order and easy access to the necessary tools, invoices, and other equipment, the processes/operations that we are responsible for cannot be conducted efficiently. A mechanic who spends ten minutes looking for a tool to repair a piece of equipment has wasted time. A manager who can't locate a prospect's contact information has lost a sale. A crew of workers that leaves a 21" mower and a weed eater at the shop has disrupted the lawn maintenance process and added significant non-value-added time to the job. A three-week delay in processing client invoices impacts cash flow and raises the question, "What is the problem with accounts receivable"?

Creating a lean culture

So once again we focus on eliminating waste. The 5-S methodology of lean management enables employees to achieve this goal by following the "S" sequence listed below:

1. **Sort:** Remove any items that are not essential to a process.
2. **Straighten (set in order):** Organize essential items and label, or outline their locations.
3. **Shine:** Clean the area.
4. **Standardize:** Establish procedures, communicate them in writing, posting where appropriate for maintaining the first 3 S's.
5. **Sustain:** Conduct periodic 5-S audits to sustain the process and maintain its presence as a catalyst to the lean management process.

Implementing the **5-S process** – sort, straighten, shine, standardize, sustain – lays the foundation for constructing a lean management culture. The process is perpetuated on a scheduled basis. Ariens, a manufacturer of lawn mowers and snow blowers, uses 5-S audit sheets for each area within their production plant (Figure 11.3). Each one of the S's is given a score, with the violations noted on a tally sheet (Figure 11.4). The employees in each area have the responsibility as a team to correct the violations prior to the next 5-S audit.

THE 5-S AUDIT CHECKLIST

Department/Area: _____ Date: _____ Person: _____

SORT (SEURU): Distinguishing between the needed and the not needed — Score

0. No activity has occurred.
1. The criteria to determine needed vs. unneeded items has been determined. Needed items for the work area have been identified. There are no more than **8** violations of unneeded items present in work area.
2. Initial red tagging exercise has been conducted and unneeded items have been removed. There are no more than **6** violations of unneeded items present in work area.
3. A separate red tag holding area exists and an uneeded items log is posted in the plant. There are no more than **4** violations of unneeded items in work area.
4. Red tagging is performed at set time intervals, and the red tag holding area is evaluated at set time intervals. All items are reviewed regularly for need. There are no more than **2** violations of unneeded items in the work area.
5. Only needed items ever enter the work area. There are **no** violations of unneeded items in the work area.

SET IN ORDER (SEITON): Organizing for ease of use, straightening up and putting things away — Score

0. No activity has occurred.
1. All needed items are present, it's not difficult to determine items in use. There are no more than **8** set-in-order violations.
2. It is obvious where needed items (including tooling, tools, procedures, etc.) belong (using lines, labels, signs). There are no more than **6** set-in-order violations.
3. The entire work area is visually indicated (including aisleways, workstations, equipment, storage locations, etc.) There are no more than **4** set-in-order violations.
4. Items are put away immediately after use. It is easy to determine what items are in use. There are no more than **2** set-in-order violations.
5. Height and quantity limits are visually obvious. There are **no** set-in-order violations.

Figure 11.3 Audit Check List
Source: Ariens Company, Brillion, WI

SHINE (SEISO): Sweeping, scrubbing and cleaning and keeping things that way	Score
0 No activity has occurred. 1 Area cleaning is done randomly. There are no more than **8** cleanliness violations. 2 Initial cleaning has occurred (floors, walls, stairs, surfaces). Machines and equipment have been cleaned. There are no more than **6** cleanliness violations. 3 Cleaning/housekeeping responsibilities are documented and followed daily. Cleaning materials are easlily accessible. There are no more than **4** cleanliness violations. 4 Cleaning is used as an inspection tool for preventative maintenance. Cleanliness problems are identified and preventative action is taken. Machines have been painted. There are no more than **2** cleanliness violations. 5 The entire work area is spotless. Surgery could be performed in the area. There are no cleanliness violations.	

STANDARDIZE (SEIKETSU): Implementing standard procedures for the first 3-S's and visual controls	Score
0 No activity has occurred. 1 5-S standards for conditions of Sort, Set in Order and Shine have been set. **Each of the first three S's is rated 1 or higher.** There are no more than **8** standardize violations. 2 5-S standards are documented and posted in work area using a workplace scan display or other visual method. **Each of the first three S's is rated 2 or higher.** There are no more than **6** standardize violations. 3 Needed items, Standard Work for 5-S and Visual Controls are in work area. **Each of the first three S's is rated 3 or higher.** There are no more than **4** standardize violations. 4 5-S is measured and posted in work area. **Each of the first three S's is rated 4 or higher.** There are no more than **2** standardize violations. 5 There are no standardize violations.	

SUSTAIN (SHITSUKE): Establishing the necessary discipline to consistently sustain 5-S stds.	Score
0 No activity has occurred. 1 Employees in work area follow 5-S guidlines. **Eash of the first 4S's is rated 1 or higher.** There are no more than **8** sustain violations. 2 Employees in work area follow 5-S guidlines and tasks are performed on a routine basis. **Each of the first 4 S's is rated 1 or higher.** There are no more than **6** sustain violations. 3 Employees in work area perform daily and/or weekly 5-S activities as part of their standard work. **Each of the first 4 S's is rated 1 or higher.** There are no more than **4** sustain violations. 4 Employees in work area perform 5-S standard work and track performance measures. **Each of the first 4 S's is rated 4 or higher.** There are no more than **2** sustain violations. 5 Area employees help create a planning worksheet to sustain 5-S standards and guidelines. 5-S activities are documented in each employee's standard work instructions. **Each of the first four S's is rated 5.** There are no sustain violations.	

	5-S Total	

Figure 11.3 (Continued)

5-S Violations Tally Sheet

The following are examples of violations to look for when conducting a 5-S audit. Observe the entire work area, and note each occurrence of a violation in its corresponding row. Explain the location of each violation and then total the number of violations for each of the 5-S's per work cell. General guidelines for defining work cells:

1. If plan is cellular, use cells or grouping of small cells.
2. If plan is functionally organized (Lathe Dept., Press Room, etc.), group the machines into a reasonable number (3-7 machines).
3. If plant is process organized (plating line, heat treat line, etc.), break the area into logical sub-areas (e.g. pre-treat, treat, post-treat).
4. If plant has large flow lines or cells (Door assembly), break it into sub-sections at logical break points. Small flow lines would fall into the cellular category.

SORT (SEIRI): Distinguish between the needed and the not needed	Totals
Unneeded equipment, tools, furniture, etc. are present.	
Unneeded items are on walls, bulletin boards, etc.	
Items are present in aisleways, stairways, corners, behind machines, etc.	
Unneeded inventory, supplies, parts, or materials are present.	
Personal belongings are present (food, jackets, etc.)	
Safety hazards (water, oil, chemicals, machines) exist.	
SET IN ORDER (SEITON): Organizing for ease of use, straightening up and putting things away	**Totals**
Correct places for items are not obvious.	
Items are not in their correct places.	
Aisleways, workstations, equipment locations are not indicated.	
Material locations and required quantities are not indicated.	
Items are set on machines/cabinets, items are leaning against walls.	
Items cannot be located within 10 seconds.	
SHINE (SEISO): Sweeping, scrubbing and cleaning and keeping things that way	**Totals**
Floors, walls, stairs, and surfaces are not free of dirt, oil, and grease.	
Equipment is not kept clean and free of dirt, oil, and grease.	
Absorbent materials used to keep oils under control. Chips fall randomly.	
Lines, labels, signs, etc. are not clean and unbroken.	
Area lacks adequate lighting, ventilation or is not free of dust/odors.	
Any other cleaning problems are present.	

Figure 11.4 5-S Violations Tally Sheet
Source: Adapted from forms used by Ariens Company, Brillion, WI

Creating a lean culture

STANDARDIZE (SEIKETSU): Implementing std. procedures for the first 3-S's and vis. controls	Totals
Necessary information/standards not known or visible.	
Out of date, torn or soiled documents are displayed in your area.	
Machine controls or critical maintenance points (fluid and lube levels) not labeled.	
Checklists not posted for all cleaning and maintenance jobs.	
Documents not clearly labeled (binders, books, etc.)	
SUSTAIN (SHITSUKE): Establishing the necessary discipline to consistently sustain 5-S stds.	Totals
How many times last week was prescribed 5-S standard work not performed?	
How many times last week were required maintenance activities not performed?	
How many times were 5-S audits not performed at the set time intervals?	

Figure 11.4   (Continued)

5-S violations tally sheet

The following are examples of violations to look for when conducting a 5-S audit. Observe the entire work area and note each occurrence of a violation in its corresponding row. Explain the location of each violation and then total the number of violations for each of the 5 S's per work cell. General guidelines for defining tools and equipment storage:

1. Outline designated areas for specific size mowers
2. Designate areas for work order repairs of mowers and equipment
3. Allocate areas for backup equipment inventory with the protocol posted for checkout
4. Clearly define an organized parts inventory section with protocol posted for checkout

KAIZEN EVENT

Kaizen (kie-zen) is a Japanese term that is defined as continuous incremental improvement of an activity to create more value and reduce **muda** (waste).

The road to improvement lies with a concerted effort of employee teams who analyze and address operational problems with corrective actions. This effort is concentrated in a five-day kaizen event. Teams of three to eight individuals consist of members of management together with selected production personnel (Figure 11.5). The initial investment in planning time prior to the event as well as afterward provides long-term benefits by assessing the value added for the customer.

Creating a lean culture

PROJECT COVER SHEET

☐ Flow ☐ Standard Work ☐ Setup ☐ Quality Other
        ☐ 3P   ☐ TPM   ☐ Ergo   VSM

Project Start Date: _____ Location: _____
Duration Of Event: _____ Value Stream: _____ Consultant: _____

Project Discription:

Why are we doing this project:

Role	Team Member Name	Home Location
Leader		
Co-Leader		
Member		
Member		
Member		
Member		
Member		
Member		

Figure 11.5 Project Team
Source: Ariens Company, Brillion, WI

Over the five-day event, the team follows the sequence below:

Day one: identify and document the problem

- Visit the site of the operation and observe the problem.
- Ask the why questions, directed toward determining the cause of the problem.
- Videotape and time operators through all phases of their work.

Creating a lean culture

- Diagram the pattern of movement from start to finish and provide quantitative data in the form of time and distance measurements (Figure 11.6).
- Graphically illustrate the information from every phase of the operation in a process map.
- Identify problems and propose seven solutions to resolve them.
- Review the day's observations and identify problems and non-valued-added time (waste). Classify areas of waste and propose alternatives, which will be tested on day two.

TIME OBSERVATION FORM

Process for Observation:											Date & Time	
Observer:											Team #:	

Seq. No.	Component Task	VA	NVA	1	2	3	4	5	6	7	8	9	10	Best Time	Points Observed
1															
2															
3															
4															
5															
6															
7															
8															
9															
10															
11															
12															
13															
14															
15															
16															
17															
18															
19															
20															
21															
22															
23															
	Time for 1 Cycle														

Best time is the lowest observed time unless there is documented valid reason that time could not be repeated. Examples: the work was done out of order, the operator had sub-assembled ahead, someone helped.

Figure 11.6 Time Observations
Source: Ariens Company, Brillion, WI

Creating a lean culture

Day two: test alternatives

- Interface with operators and discuss proposed alternative measures for waste reduction.
- Oversee "**try-storming**," which tests the alternative measures proposed for the operation in a succession of field trials.
- Integrate methods that improve the operation.
- Observe the operation with integration of the improved methods and establish new standards.

Day three: goal improvement assessment

- Confer with operators on the impact of improved operation methods on the resolution of operation problems.
- Propose additional alternatives if some problems have not been addressed.
- Quantify alternatives with time trials for the final assessment of improvements.
- Videotape the operation to document improvements and use as a visual record of performance standards to train employees.

Day four: develop a standard working document

- Observe improved operation and break it down into a step sequence.
- Describe each step in detail with new standard practices.
- Diagram work site with labeled designations of each step.
- Retest the improved operation with an experienced employee.
- Demonstrate the improved operation to all employees who will be engaged in its implementation.

Day five: celebrate

- Present all of the team's accomplishments and success.
- Acknowledge all team members and employees involved in the kaizen event.
- Present a summary of "Lessons Learned" (Figure 11.7).
- Summarize efforts toward increasing operation efficiency and bottom-line profitability.

One of the important aspects of lean management culture is communication of the successes with all employees. This is accomplished by posting written indicator reports, e.g., target sheets (Figure 11.8) and a kaizen event newspaper (Figure 11.9). These visuals affirm the success of the principles of lean culture in enhancing the profitability of the company through constant improvement. Most importantly, it engages everyone in the company and lets them know how well they are performing as a team.

Lessons Learned: _____

_____

_____

_____

_____

_____

_____

Thanks To: _____

_____

_____

_____

_____

_____

_____

Figure 11.7 Lessons Learned

Source: Ariens Company, Brillion, WI

KAIZEN EVENT TARGET SHEET

Focused Area: _____   Event Start Date: _____

	Description of Targets	Start	Target	Day 1	Day 2	Day 3	Day 4	% Increase/ Decrease
1	Space (Sq. Ft.)							
2	Inventory (pcs)							
3	6S Scroe							
4	Safety Score							
5	Takt Time (seconds)							
6	Throughput							
7	Volume per Day							
8	Full-time Equivalent Crew							
9	Productivity (units/Emp. Hr.)							
10	Schedule Attainment							
11	Total Kaizen Implements							
12	Safety Kaizen Implements							
13	Quality Kaizen Implements							

Remarks: _____

_____

_____

Figure 11.8 Target Sheets

Source: Ariens Company, Brillion, WI

KAIZEN NEWSPAPER

Page: _____ of: _____

Project Description: _____

Date: _____

No.	Description of Problem	Counter Measure	Person Responsible	Due Date	% Complete	Date Complete
1					25% / 50% / 75% / 100%	
2					25% / 50% / 75% / 100%	
3					25% / 50% / 75% / 100%	
4					25% / 50% / 75% / 100%	
5					25% / 50% / 75% / 100%	
6					25% / 50% / 75% / 100%	
7					25% / 50% / 75% / 100%	
8					25% / 50% / 75% / 100%	
9					25% / 50% / 75% / 100%	
10					25% / 50% / 75% / 100%	
11					25% / 50% / 75% / 100%	
12					25% / 50% / 75% / 100%	
13					25% / 50% / 75% / 100%	
14					25% / 50% / 75% / 100%	
15					25% / 50% / 75% / 100%	
16					25% / 50% / 75% / 100%	
17					25% / 50% / 75% / 100%	
18					25% / 50% / 75% / 100%	
19					25% / 50% / 75% / 100%	
20					25% / 50% / 75% / 100%	

Figure 11.9 Kaizen Event Newspaper
Source: Ariens Company, Brillion, WI

Creating a lean culture

QUANTITATIVE INDICATORS OF LEAN MANAGEMENT

The indicators described throughout this chapter further reinforce the statement from Chapter 5: "You are what you measure." In this application, measuring the current lean status of a company means using quantitative measurements to reflect the company's progress.

Sales per dollar of salaries

Quantitative indicators of lean management that can be included in the visual postings are the **ratio of sales per dollar of salaries**, which represents the dollar amount of sales that are generated by each dollar of salaries. A company with gross sales of $2.5 million and salaries of $800,000 has a ratio of 3.2, indicating that $3.20 of sales are generated by each dollar of salaries. Once a baseline ratio is established, each subsequent year can be evaluated by comparison. An increase in sales the subsequent year to $3 million with an increase in salaries to $850,000 increases the sales-to-salary ratio to 3.5, which is a 9% improvement. However, if salaries had increased to $1.2 million while sales increased to $3 million, the ratio would be 2.5, which represents a 28% decline. This may be an indication of either being top heavy in management salaries or incurring waste associated with non-value-added time.

Profit per dollar of sales

Another ratio measurement that serves as a lean indicator is the **ratio of profit per dollar of sales**, an indicator of cost reductions and effective waste management. A company with sales of $1.7 million and profit of $255,000 would have a ratio of 15% profit for every dollar of sales. If its sales increased the following year to $2.25 million and its profit decreased to $180,000, the ratio of profit to sales would be 8%, indicating an increase in non-value-added time.

Lean management implementation in a business operation requires outside consultants to work with management and train them on the processes associated with developing a lean culture. For further information, see the section Lean Management Resources at the end of the chapter.

CASE STUDY

NATIONAL ASSOCIATION OF LANDSCAPE PROFESSIONALS
(FORMERLY PROFESSIONAL LANDCARE NETWORK)

CRYSTAL BALL REPORT #26

DEICING EQUIPMENT INSTALLATION & REMOVAL

Creating a lean culture

Identifying the problem

Landscape companies that also provide snow removal and deicing services are often confronted with an overlap of seasons. This presents a problem when their trucks are needed for early spring cleanups while salt spreaders are mounted on the back of the trucks, since the spreaders are still needed at times during this transition period. The changeover of equipment is a non-value-added, time-consuming operation.

Total changeover operation (14 salt spreaders – 5 large, 9 small):

Installation:	15 hours, 30 minutes
Removal:	10 hours, 5 minutes
Crew members:	3–4
Tools:	12

Project justification

If the changeover operation time could be reduced then there would be more value-added time in delivering spring cleanup and deicing services. In addition, redirecting mechanics' time to providing services instead of spending a few days changing over equipment would also represent a potentially significant savings in labor costs.

> **Goal:** Reduce total changeover operation to four hours for salt spreader installation and eight hours for removal.

Team members

- Lean management facilitator
- Fleet and facilities manager
- Lead mechanic
- Account manager
- Administration representative
- Sales representative

The diverse team provided an opportunity for different perspectives and challenges to the current operation.

Creating a lean culture

Observing and documenting the problem

The team was briefed on the operation and the entire changeover operation process was observed and timed. Based upon the observations, the why questions were posed, followed by a discussion of alternatives to the current steps of the operation.

Documentation

Installation:	Large salt spreader: 1 hr 45 min
	Small salt spreader: 45 min
Removal:	Large salt spreader: 42 min
	Small salt spreader: 39 min
Total installation (14 units):	15 hr 30 min
Total removal (14 units):	9 hr 21 min
Crew members:	3–4
Tools:	12

Goal improvement assessment

Process improvements were tested for both size spreaders and time measurements were recorded. The changeover improvements entailed retrofitting the salt spreaders and truck beds with welded components for forklift installation and removal.

Assessment

Installation	Large spreaders: 9 min 36 sec = 9.6 min
	Small spreaders: 4 min 23 sec = 4.38 min
Removal	Large spreaders: 5 min 45 sec = 5.75 min
	Small spreaders: 2 min 12 sec = 2.2 min
Crew members:	2
Tools:	1

Standardize operation

The following improvements were implemented into the salt spreader changeover operation:

- Placement positions were labeled on all sides of the salt spreaders
- Skid systems were welded on the bottom of salt spreaders to slide into truck bed

Creating a lean culture

- Forklift pockets were welded on sides of spreaders
- Alignment racks were welded into the truck beds
- Stop bars were welded to the bottom of the spreaders
- Drop chutes were permanently attached to the spreaders
- Chains were installed to stabilize removed spreaders in the truck bed

Celebrate

The results were presented and the team acknowledged for exceeding the kaizen event goal.

Kaizen event results

Total installation

Large spreaders: 48 min (5 × 9.6 min)
Small spreaders: 39.42 min (9 × 4.38 min)

Total removal

Large spreaders: 28.75 min (5 × 5.75 min)
Small spreaders: 19.8 min (9 × 2.2 min)

New operation status

Total installation: 87.42 min or 1 hr 17 min 25 sec
Formerly   930 min (15 hr 30 min)
                    amount of downtime
                    reduced by 90.6%
Total removal: 48.55 min or 48 min 33 sec
Formerly   561 min (9 hr 21 min)
                    amount of downtime
                    reduced by 91.3%

Creating a lean culture

SUMMARY

What is the philosophy of lean management?

The lean management philosophy originated with the Toyota Production System (TPS), which was developed by a Toyota engineer, Taiichi Ohno. It is a long-term commitment to continually strive to provide value to the customer. Incremental improvement of the company's operations minimizes non-value-added time and increases value-added time to eliminate waste. All the company's employees and business partners also are engaged to work together as a team to achieve improvement. The **4 P's** of TPS are listed below:

1   **Philosophy:** a commitment to function as a complete entity that will strive to constantly identify problems, with a focus on reducing waste to provide value to the customer.
2   **Process:** Eliminate waste throughout the company's operations.
3   **People and partners:** Develop a lean culture that gets employees and suppliers to buy in to lean processes. Engage these individuals to work together to improve the delivery of services or products to the customer.
4   **Problem solving:** Engage employees at all levels to contribute to solving operation problems and implementing working lean.

What are the principles of lean management?

The principles of lean management are embodied in its philosophy to develop a culture that will focus on reducing non-value-added time and increasing value-added time. Decisions are made from the bottom up and leaders emerge who perpetuate the culture by serving as mentors to new employees. The lean culture communicates the waste and inefficiencies and tracks progress through visuals as they are addressed.

What is a lean culture?

A lean culture manifests a company's lean management philosophy of commitment to incrementally improving its functions to provide value to its customers. This entails a process that is focused on reducing waste by constantly striving to identify inefficiencies and implement improvements throughout the company's operations. This starts with the 5 S's – sort, straighten, shine, standardize, and sustain – and is followed by kaizen events to actually work out how to improve operations.

Creating a lean culture

The lean culture engages all of its personnel and associated business partners in decision making and acknowledges all those who contribute to process improvement. The lean culture is never satisfied with the status quo, believing there is always room for improvement. "Best is the worst enemy of better." Therefore, a lean culture is always confronting problems and coming up with solutions.

What are kaizen events?

Kaizen events are conducted over a five-day period with the primary objective of identifying operational problems and implementing incremental improvements. The kaizen team is comprised of management and select production personnel.

During the kaizen event, team members observe, identify, and document existing problems prior to proposing alternatives. The proposed alternatives are discussed, tested, and measured. Based upon the assessment of the tests, appropriate improvements are implemented and standardized into the operation.

KNOWLEDGE APPLICATIONS

1. Observe the operations of a fast-food or other retail business and conduct an assessment of non-value-added time. Describe how value-added time could enhance the service delivered to the customer.
2. Conduct a 5-S audit of your personal space.
3. Select a day to observe and document waste in processes within your environment.
4. Do an assessment of your typical eight-hour day. Identify where there is non-value-added time (muda). How can this time be converted into value-added time?
5. Describe how you would conduct a kaizen event with your current or past working environment.

LEAN MANAGEMENT RESOURCES

Training

JP Horizons, Inc. "Working Smarter." www.jphorizons.com
   This online 52-week lean management course is part of a complete lean training program developed by JP Horizons, a business management consulting company, to engage employees at all levels.

FURTHER READING

The Goldmine: A Novel of Lean Turnaround. Freddy Balle and Michael Balle, 2005. Cambridge, MA: Lean Enterprise Institute.

The Kaizen Revolution: How to Use Kaizen Events to Double Your Profits. Michael D. Reagan, 2000. Raleigh, NC: Holden Press.

Lean Management for the Green Industry: An Operational Strategy that Delivers Value to Customers and Eliminates Waste. Crystal Ball Report #26. James P. Martin, 2006. Herndon, VA: National Association of Landscape Professionals.

The Toyota Way Fieldbook: A Practical Guide for Implementing Toyota's 4 P's. Jeffrey K. Liker and David Meier, 2005. New York and Chicago: McGraw-Hill.

NOTES

1. Dan Ariens and Ariens Company, 2006, *Crystal Ball Report #26* (National Association of Landscape Professionals [formerly Professional Landcare Network]), Herndon, VA.
2. Jeffrey K. Liker and David Meier, 2006, *The Toyota Way Fieldbook*. McGraw-Hill.
3. Dan Ariens and Ariens Company, 2006, *Crystal Ball Report #26* (National Association of Landscape Professionals [formerly Professional Landcare Network]), Herndon, VA.

12
Professional development

CHAPTER OBJECTIVES

To gain an understanding of:

1 Professional development goals
2 Professional development resources
3 Multilevel development training programs
4 Integrating professional development into the company culture

KEY TERMS

certification programs
trade associations
training investment
training programs

Whether it is referred to as training or skill development, the goal is the same: to provide employees with professional development. Professional development is applicable to all positions within a company. The landscape industry could not have evolved to its status today without educational training acquired through academic programs, through trade association seminars and workshops, and from consultants.

As companies evolved from pickup truck operations to multimillion-dollar corporations, owners recognized the necessity of instituting internal and external training programs. The impetus for expanding these programs is driven by a competitive market that necessitates maximizing the productivity potential of human assets.

Professional development

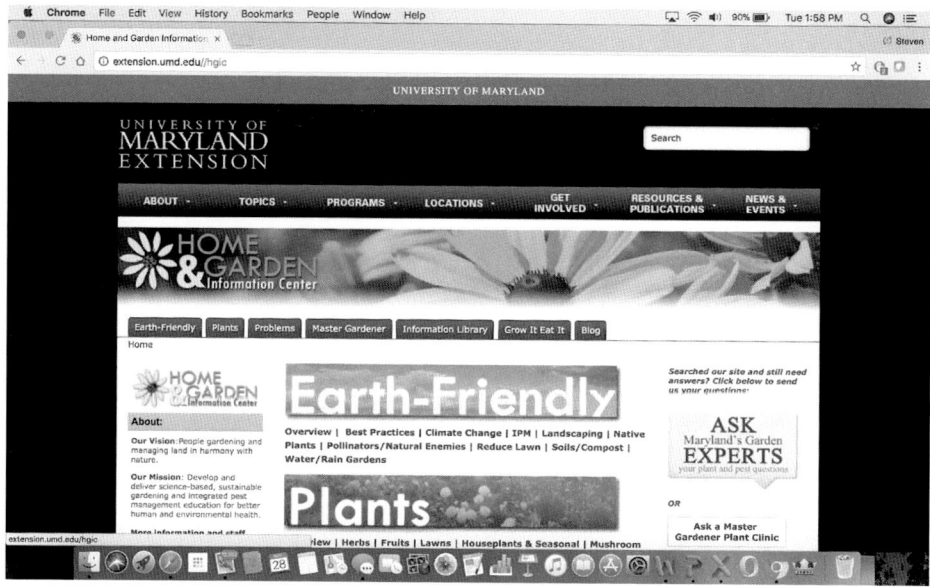

Figure 12.1 Web Access to University Agriculture Extension Sites
Source: University of Maryland Cooperative Extension, www.extension.umd.edu/hgic

Companies that succeed develop their human assets. These employees perform at high levels of competency and are proactive in improving the productivity and efficiency of their team. Professional development is an ongoing process that perpetuates a company's growth and profitability.

No one will dispute the fact that there is a direct correlation between training and productivity. Regardless of the level of one's education or professional experience, there are daily reminders that "you don't know what you don't know"[1] (Landon Reeve, Chapel Valley Landscape Company). It is therefore imperative that companies continually take the opportunity to offer their employees learning venues that will benefit their professional development. Within the landscape industry, there are several venues that provide opportunities for acquiring skills and knowledge:

Trade associations that conduct conferences and workshops

National Association of Landscape Professionals, www.landscapeprofessionals.org
Professional Grounds Management Society, www.pgms.org
American Horticulture, www.americanhort.org
International Society of Arboriculture, www.isa-arbor.com
American Society of Landscape Architects, www.asla.org

Professional development

Irrigation Association, www.irrigation.org
State and local landscape associations

Manufacturers' technical representatives that offer training

On site
Local seminars/workshops

Consulting firms

On site

Publications

Trade association journals
University Agricultural Extension bulletins
Horticulture research journals

Websites

Trade associations
University Agricultural Extension
University/college long-distance learning
Internet search engines

Courses

University Agricultural Extension
- Short courses, seminars

Universities and community colleges
- Online training
- Courses, seminars
- Consulting companies
  Trade associations
  Certifications

Professional development

External professional development resources do not replace in-house training programs, but they are an important adjunct to the whole development process. Individuals attending conferences and workshops come back inspired with new ideas and are enthusiastic to see the results of implementing these ideas. The return on investment from those attending far exceeds the investment associated with travel and conference fees. Motivation, improved job performance, and employee retention are some of the dividends derived from this form of human asset development.

TRADE ASSOCIATIONS

Trade association conferences are one of the best resources for professional development. They provide an opportunity to learn from industry leaders and consultants on subjects ranging from financial management to water management. One of the largest conferences, attended by thousands of landscape contractors, is the Green Industry Conference (GIC). The **trade associations** that coordinate the conference are the National Association of Landscape Professionals (NALP) and the Professional Grounds Management Society (PGMS). This conference is held annually in Louisville, KY, in conjunction with the Green Industry and Equipment Expo. This learning venue also provides an opportunity to network with industry peers for an exchange of ideas and mentoring. Informal breakfast and lunch meetings, as well as receptions, provide a format for exchanging information. An example of such a format is "Breakfast with Champions" at the GIC. The champions are industry leaders who address specific topics in roundtable discussion groups. Topics cover human resources, financial management, sales and marketing, leadership, design/build installation, maintenance, irrigation, and arboriculture (see Table 12.1). Daily seminars are conducted on subjects ranging from business management to storm water management systems.

Table 12.1 Examples of Conference Topics

Executive Management	Middle Management	Field Supervisors
Defining leadership	Managing a multicultural labor force	Water conservation principals
Hiring, inspiring, retaining a labor force	Profitability through productivity	Lawn pesticides
Risk management strategies	Cutting-edge sales techniques	Perennials, ornamental grasses
Communicating a business vision	GIS/GPS mapping systems	Landscape diagnostics
Open book management	Client management	Color in the landscape
Smart growth marketing		Training the trainer

Professional development

SEMINARS AND WORKSHOPS

Seminars and workshops conducted by business consultants address an array of subjects to provide personal and professional development. The workshops are structured to engage management personnel in interactive exercises so that they can apply their newly acquired knowledge and skills. Examples of workshop topics are:

- Human resource skill building: recruiting, hiring, coaching
- Just-in-time training program development
- Time management skill building
- Leadership development:
    - Goal setting
    - Team building
    - Communication
    - Planning
    - Delegating

Figure 12.2 Networking Green Industry Conference
Permission granted by the National Association of Landscape Professionals, Herndon, VA

Figure 12.3 Industry Peers Serve as Mentors
Permission granted by the National Association of Landscape Professionals, Herndon, VA

Figure 12.4 Landscape Industry Trade Shows Serve as Technical Resources
Permission granted by the National Association of Landscape Professionals

Table 12.2 JP Horizons Better Results Campaign

Campaign: 100 Viable Leads	
Specific result:	Develop a team-oriented program to find viable leads to pursue.
Dynamic action 5:	Establish 100 leads and track the business they generate. Celebrate team's achievement.
Dynamic action 4:	Assign team members to lead focus areas: former clients, drive-bys, networking, e.g., apartment associations, property management companies, homeowner associations, and existing clients.
Dynamic action 3:	Select a leader to direct the campaign. Develop a lead sheet.
Dynamic action 2:	Conduct a campaign orientation, discussing the format, process, and how to generate leads.
Dynamic action 1:	Develop an action plan that includes a lead sheet, a tracking database, and a posted campaign progress chart.

Source: "Leadership Insights," JP Horizons, Inc., Landscape Management, August 2003

Programs such as JP Horizon's Face-to-Face events train workshop participants to utilize a structured approach to attain results through the Better Results Campaign. The workshop provides a template or guide that employees can utilize to achieve specific goals. An example of an application of the Better Results Campaign is illustrated in Table 12.2. The campaign's goal is to develop 100 viable leads for expanding sales. A sales manager would engage the sales team in the action steps that would lead to the campaign's defined goal. Time lines and team member responsibilities would be established for each step of the campaign.

TRAINING PROGRAMS

Training programs are an integral part of a company's culture because they are directly responsible for increasing productivity and efficiency. Just as important as the end result of these programs is that they represent a company's commitment to its employees. A comprehensive series of training programs for all levels of employees provides them with the knowledge and skills required for advancement. The benefits to the company from its **training investment** are manifested in the employees' increased productivity, morale, empowerment, ability to make decisions, and retention. Retention of trained employees provides the company with a competitive edge, since their work manifests quality, consistency, and efficiency. A company's financial success hinges on its team of quality employees. Landscape companies generally target a 65% retention rate of field personnel, 85% of field supervisors, and 90% of managers.

Successful training programs are contingent upon the program structure and the ability and knowledge of the trainer. Trade associations and consultants offer seminars and

Professional development

workshops to develop trainers' skills. Kraft Associates/ODA Inc. conducts training seminars based on the K.A.S.H.© system: training programs that develop employees' knowledge, attitude, skills, and habits. The titles of these seminars, "Training the Trainer," have become a cliché in the industry. The common objective of these seminars is to provide participants with the skills necessary to conduct an effective training program. The skill topics encompass:

- Communication
- Organization
- Instructional materials
- Role playing
- Program evaluation
- Reinforcement

Retention of information conveyed in a training program depends on the delivery format. This point is evident by results of comprehension studies that have been done on the amount of retention from reading, hearing, seeing, speaking, and doing:

Retention rates

10% reading
20% hearing
30% seeing
50% seeing and hearing
70% speaking
90% doing

If plant identification was the training topic, a CD or web format would need to be supplemented with plant samples. That would fulfill the "doing" aspect of learning, handling and examining live plant samples. Engaging trainees in learning plant identification techniques at a job site, nursery, or arboretum would certainly reinforce the training objective.

In addition to knowledge retention, other important factors that contribute to effective training programs are:

Scheduling – convenient for the audience
Duration – the average attention span is twelve minutes, so three topics can be covered in a half-hour session
Consistency – weekly, monthly, seasonal content – emphasis on application versus theory
Bilingual – all verbal instruction and instructional materials need to address the multicultural labor force

Professional development

"Just-in-time" training – information presented on site for immediate application
Manuals – to illustrate and emphasize the topic

The infrastructure of a training program for field personnel consists of a series of customized subject modules that address topics which have a direct impact on production efficiency. Examples of training modules for landscape maintenance and installation divisions are presented in Table 12.3.

Table 12.3 Training Modules for Landscape Maintenance and Installation Divisions

	Maintenance Division			Installation Division
I	*Equipment Operation*		V	*Plant Installation*
a	Operation procedures		a	Plan/Blueprint interpretation
b	Safety		b	Plan take-offs
c	OSHA requirements		c	Specifications
d	State Department of Transportation Regulations		d	Site analysis
e	Maintenance		e	Equipment operation
f	Troubleshooting		f	Safety
			g	Site preparation
II	*Turf Management*		h	Planting procedures
a	Mowing height		i	Change orders
b	Edging and trimming		j	Plant establishment procedures
c	Aerating			
d	Fertilizing		VI	*Irrigation System Installations*
e	Weed identification and control		a	Design interpretation
f	Pest and disease control		b	Site analysis
g	Diagnostic skills		c	Equipment operation
h	Irrigation system maintenance		d	Safety
			e	Manifold installation
III	*Shrub Management*		f	Pipe and sprinkler installation
a	Pruning Techniques		g	Controller installation
b	Plant identification		h	Controller programming
c	Pest and disease control		i	System troubleshooting and maintenance
d	Bed edging		j	Winterizing system
e	Weed control		k	Spring startup procedures
IV	*Pesticide Application*			
a	Material safety data sheet			
b	Environmental Protection Agency requirements			
c	State agricultural agency regulations			
d	Equipment calibration			
e	Pesticide mixing procedures			
f	Application techniques			
g	Sprayer cleaning procedures			
h	Chemical spectrum of activity			
i	Phytotoxicity symptoms			
j	Diagnostic skills			

Professional development

Many topics can be introduced into a company's training program, some of which can be delivered in daily "just-in-time" situations and others over longer periods of time. Some companies have incorporated online operation management training for entry-level managers. Incorporated throughout the training programs are the company's standards and its commitment to constantly improve its services through the professional development of its employees.

Regardless of the level of training, the trainer's mode of information dissemination is directly related to the comprehension and retention of the trainees:

"Tell me – I'll forget."
"Show me – I'll remember."
"Involve me – I'll understand."[2]

These statements reiterate the previously mentioned retention rates, showing that an individual's act of doing reinforces the learning process. They also add another dimension to the act of doing: comprehension.

The training process doesn't stop at the end of the session. Monitoring, evaluation, and reinforcement are also part of the process. These phases of the process assess the impact of the program and determine whether supplemental training is necessary. The "training loop" in Table 12.4 represents the phases associated with the training program cycle.

Table 12.4 The Training Loop

Input	Action	Output
Training plan Scheduling	Training program	Production
Reinforce/Modify	*Evaluate*	*Feedback*
Review training Modify training	Results: expected versus actual Impact Clients Employees Company	Measure productivity Quantitative Qualitative

Source: K.A.S.H.© A System of Training, Kraft Associates/ODA, Inc.

Professional development

CERTIFICATION PROGRAMS

Certification programs have been developed by national and state trade associations to elevate the professional standards in their fields. Training materials are available through these associations, and they offer regional seminars and testing sites. Landscape companies use these programs to advance the knowledge and skills of their employees. Attaining certification is regarded as a significant step in an employee's professional development and most often is tied to financial incentives. In addition, a company has a competitive edge by having certified employees. A company will acknowledge these certifications in their proposals to prospective clients.

The National Association of Landscape Professionals administers six Landscape Industry Certifications:

LIC Manager
LIC Technician – Interior
LIC Technician – Exterior
LIC Lawn Care Manager
LIC Lawn Care Technician
LIC Horticultural Technician

LANDSCAPE INDUSTRY CERTIFIED TECHNICIAN – EXTERIOR

Candidates select from the following modules: Hardscape Installation, Softscape Installation, Turf Maintenance, Ornamental Maintenance, or Irrigation. Written tests and skill applications are incorporated into all of the modules.

Problem solving

First aid and safety	Irrigation controller programming
Reading landscape plans	Plant identification
Sod installation	Work orders and reports
Irrigation component identification	Truck and trailer operation
Tree planting, staking, guying	Chainsaw operation
Hands-on modules*	

Professional development

Installation	Maintenance	Irrigation
Plant layout	Pruning	Reading plans
Grading and drainage	21" mower operation	Lateral installation
Survey instrument	Walk-behind mower operation	Main line installation
Paver installation	Riding mower operation	Valve repair
Rototiller operation	Pesticides	Valve wiring
Tractor/skid steer Operation	Fertilizers	
Aerator operation	Pipe installation – trencher	
Blower operation	Pipe installation – puller	

*Each module also has a written comprehension component.

LANDSCAPE INDUSTRY CERTIFIED MANAGER

The LIC Manager exam is restricted to candidates who are business owners or managers. Its development resulted from a collaborative effort between the National Association of Landscape Professionals and the Professional Development Institute. Certification candidates are tested on subjects associated with business management.

Business planning	Sales
Accounting	Marketing
Management	Communications
Liability risk	Public relations
Contract law	Human resources
Safety	Production operations

    Training is an ongoing process at all levels of employment. The success of training programs is contingent upon the commitment of owners and managers. This commitment should be based on the assumption that people want to excel at what they do. Developing employees through such programs improves production quality and efficiency and provides individuals with job satisfaction.

SUMMARY

Training adds value to a company's human assets. Providing structured programs develops employees' skills and provides them with challenges and increased self-esteem. Production efficiency and quality is enhanced and perpetuated at all levels.

Professional development

What are professional development goals?

Professional development goals are those that are directed toward maximizing the achievement potential of a company's human assets. These goals should be inclusive of all employees from upper management to field personnel.

What are some professional development resources?

Sources for external training venues include:

- Trade association workshops and seminars
- Certification programs
- University extension workshops and courses
- Trade conferences

In-house training programs incorporate just-in-time training for on-site instruction and implementation of production practices. The key to the success of these programs is structure, consistency, and duration. The infrastructure of these programs consists of training modules, e.g., on equipment safety, turf management, pruning, and so on. In-house training resources can be supplemented by outside technical representatives.

What are multilevel development training programs?

Multilevel training programs address the skill levels required for advancement to supervisory and management positions. These programs may require outside resources, e.g., leadership workshops, financial management seminars, and possibly bilingual communication courses at community colleges or universities.

How can professional development be integrated into a company's culture?

Professional development should be a component of every company's culture, since it is one of the primary factors associated with employee retention and productivity. Management can establish mentors for just-in-time training and provide opportunities for individuals to participate in external training and certification programs. Mentors within companies can be

Professional development

designated for areas within their expertise. On-site training sessions can also be conducted with industry consultants and manufacture representatives.

A company's investment in training programs provides dividends in the form of employee empowerment, job satisfaction, retention, and production efficiency.

KNOWLEDGE APPLICATION

1. As an entry-level landscape manager, what type of training would you expect to receive?
2. Structure a just-in-time landscape training module for an installation or maintenance operation.
3. List six professional development resources offered by one or more of the following trade associations:
   National Association of Landscape Professionals, www.landscapeprofessionals.org
   International Society of Arboriculture, www.isa-arbor.org
   American Nursery and Landscape Association, www.anla.org
   International Society of Arborists, www.isa-arbor.com
4. Review the following website for horticultural information that has application to just-in-time training of landscape field personnel: extension.umd.edu/hgic, click on the "Problems" tab. List training topics that could be developed from this information.

NOTES

1. Land Reeve, 2005, Chairman of the Board, Chapel Valley Landscape, Personal Communication.
2. Chinese Proverb.

13

Bottom-line leadership

CHAPTER OBJECTIVES

To gain an understanding of:

1. Positive leadership traits
   a. Motivator
   b. Caring
   c. Listener
   d. Communicator
   e. Mentor
   f. Team oriented
2. The effect of leadership on employee motivation
   a. Increased productivity
   b. Positive company culture
   c. Increased profitability
   d. Employee retention
3. Leadership's role in employee development
   a. Sets high expectations
   b. Empowers employees
   c. Provides development resources
   d. Engages employees at all levels
   e. Mentorship

Bottom-line leadership

When all is said and done, leadership stands out as being the one determining factor in a company's financial success. It all begins at the top with executive management's vision and structuring of the company. Once the company's infrastructure is defined, it will only move forward in a competitive market with strong leadership. There are examples in the landscape industry of those who have led and those who haven't. Companies that personify strong leadership have grown from a small cadre of employees to hundreds and from revenue of thousands to millions of dollars. The former Brickman Group (currently BrightView) started out in the Chicago area as a design/build company and subsequently developed into the largest landscape maintenance company in the country, with an annual revenue of $900 million. The former Valleycrest Landscape Company (currently BrightView), began as a nursery/landscape business in southern California and subsequently became the largest landscape construction company in the country with an annual revenue of $800 million ($1.5 billion with maintenance division revenue). This type of growth is not just related to an infusion of capital; it is directly related to leadership.

Throughout this book, the human factor has been defined as the key to success in the landscape industry. Equipment doesn't get the job done, people do. Equipment doesn't produce a profit, people do. The motivating factor that is the catalyst for production and profit is leadership – leadership that permeates the entire company.

Motivated people move faster.[1]

A company's culture evolves from the president of the company. Positive company cultures create a work environment of motivated and productive employees. These are employees who wake up in the morning and look forward to going to work. Companies with this type of culture thrive because of their level of productivity and efficiency. Their success comes from retaining employees whose experience and leadership are constantly improving the company's bottom line.

Financial management builds and maintains the framework for the company's production. It tracks production history and establishes parameters to ensure that current and future production achieves profit goals. This aspect of management is critical to the survival and future growth of all companies. However, regardless of the degree of financial management sophistication, it's production efficiency and production volume that generate black ink on the bottom line. These two production factors are prevalent in companies with motivated employees.

RESULTS-ORIENTED LEADERSHIP

Two books that are must reads on the subject of leadership are *Contented Cows Give Better Milk*, by Bill Catlette and Richard Hadden, and *Fail-Safe Leadership*, by Linda Martin and Dr. David G. Mutchler. They both clearly define leadership and relate its impact on the business world. Both books address the need to create a working environment that motivates and meets the needs of employees.

For ten years, profitability and revenue comparisons were made with "contented cow companies" and their industry peers. In all instances, the "contented" companies were consistent in profitability and exceeded the revenue of their industry rivals. Among the top 50 landscape companies that appear annually in *Landscape Management*, a trade journal, how many hold their ranking from year to year? The answer: Most of them. Does this mean that they are "contented cow" companies? There certainly must be some level of contentment in these organizations that enables them to attain their level of sales revenue. The best way to retain a loyal customer base is through establishment of a fully engaged workforce.

The results that every company is seeking are increased sales, productivity efficiency, and profitability. All of these results are people driven and strongly dependent on leadership. This requires leadership that:

- Motivates employees to work faster and more efficiently
- Sets high expectations

   High expectations breed high performance.[2]

- Empowers employees by providing training resources

   An individual without information cannot take responsibility.
   An individual who is given responsibility cannot help [but take responsibility].[3]

- Listens to its people
- Cares about its people

   If you care you're there.[4]

- Develops people
- Engages employees in operations management

   Contented people give better performance.[5]

COMPANY VISION AND MISSION

"Contented companies" not only meet the leadership criteria previously stated, but most importantly, they engage every employee in fulfilling the company's vision. This vision is expressed in a vision statement that describes the future, where the company/organization is going, or where it wants to go. Amazon's statement is "Our vision is to be earth's most customer centric company, to build a place where people can come find and discover anything that they might want to buy on line."[6] Vision statements provide a foundation for employee commitment and understanding of management's perspective of the future. They should project an image of the future and be both dynamic and innovative.

Mission statements focus attention on essentials and summarize the core competencies and capabilities of the business. A mission statement expresses the basis for the company's existence, who it is and its core values. An excellent example of such a statement is that of the American Red Cross:

> The American Red Cross is a humanitarian organization, led by volunteers, that provides relief to victims of disasters and helps people prevent, prepare for, and respond to emergencies. It does this through services that are consistent with its Congressional Charter and the principles of the International Red Cross Movement.[7]

Vision statements describe a company's future direction.
Mission statements summarize a company's core values and why it exists.

Together, mission and vision statements provide direction for a business by focusing attention on doing things day by day to accomplish the mission while taking steps to pursue the future vision.

Everyone is on the same page and knows where the company is going. Once everyone knows the path the company is traveling, they can determine what they can do to help it reach its destination. Mission statements foster a team-oriented environment, enhance employee morale, and provide everyone with a clear company image.

A company's leadership is a sustaining competitive advantage – leadership that knows how to set goals and achieve desired results. Leaders are guides showing the way while simultaneously directing employee performance that is necessary to reach the company's destination. These leaders set high expectations that result in high performance. They make people feel powerful rather than making them feel powerless. This goes along with having high expectations and assigning people responsibilities that instill confidence in them. Individuals on the receiving end will have the desire to prove that having faith in them is well founded.

MULTILEVEL LEADERSHIP

One primary characteristic of effective leadership is knowing when to lead and when to follow. Once the direction is established, effective leaders develop their people, empower them with the necessary resources, and motivate them to achieve to their highest level. The direction is defined in terms of the desired results, for example sod installation within a specified number of hours and to installation specifications. In addition, effective leaders develop the processes that are necessary to ensure the desired results. In the landscape industry, this would apply to operating procedures and standards, such as for tree installation and staking. Effective leaders do not micromanage, nor do they suppress innovation and problem solving at any level.

Leadership is all about results, and development is about creating a working environment in which all individuals recognize their role and responsibility in achieving the goals to which they commit. People must be empowered with processes that allow them to lead. When this is accomplished, everyone in the organization becomes a leader. Multilevel leadership results from empowerment, providing a company with decision-making depth. Just like a basketball team with a strong bench, each player knows what it takes to win the game. This team of leaders will be motivated to set positive examples, serve as mentors, and perpetuate a team culture.

Two great leaders expressed their perspective of leadership in a very similar vein:

Don't tell people how to do things, tell them what to do and let them surprise you with their results.[8]

The best executive is the one who has sense enough to pick good men* to do what he wants done, and self-restraint to keep from meddling with them while they do it.[9]

*Today, Teddy would have referred to "people."

SUMMARY

Leaders have the responsibility of developing and motivating individuals to perform to the best of their ability. Leaders influence the working environment and are responsible for motivating individuals to accomplish their goal in the most efficient manner. When this is achieved, a company has a competitive advantage, because "contented cows give better milk."[10]

Production quality and quantity are increased by leadership that:

- Motivates individuals to become more efficient in their jobs
- Sets high expectations
- Supports rather than impedes
- Listens
- Empowers individuals by providing training and information resources
- Cares about its people: "Lead by your heart"

Bottom-line leadership

- Develops leaders
- Engages employees in operations management
- Understands that positive results come from positive leadership

KNOWLEDGE APPLICATION

1. Select a landscape operation and define how multilevel leadership can enhance attainment of the desired results.
2. What skill development would strengthen your leadership qualities?
3. What impact does leadership have on the work environment and the end results?
4. Describe an existing company that personifies innovative leadership.

NOTES

1. Jim Barksdale, former CEO of Netscape.
2. Jan Carlzon, former CEO Scandinavian Airlines System.
3. Ibid.
4. Ibid.
5. Tommy Lasorda, former manager Los Angeles Dodgers.
6. Amazon Mission Statement.
7. American Red Cross Mission Statement.
8. George S. Patton, World War II General.
9. Theodore Roosevelt, 26th United States of America President.
10. Richard Hadden and Bill Catlette, 2001, *Contented Cows Give Better Milk*. Hoboken, NJ: John Wiley and Sons.

14

Economic sustainability

CHAPTER OBJECTIVES

To gain an understanding of:

1. The ramifications of a recessionary economy
   a. Lower profit margins
   b. Increased competition
   c. Reduced client budgets
2. Impacts on landscape company divisions
   a. Decreased design/build residential contracts
   b. Decreased enhancement budgets
   c. Decreased commercial construction contracts
   d. Decreased maintenance budgets
3. How to compete in the current business environment
   a. Reduce indirect and overhead expenses
   b. Increase production efficiency
   c. Reduce nonproductive time
   d. Increase client relations
   e. Diversify landscape services
4. Marketing strategies
   a. Regional branding
   b. Emphasize company attributes
   c. Value engineering
   d. Increase sales
   e. Community networking
   f. Increase advertising budgets

Economic sustainability

KEY TERMS

branding
furloughs
multiple equivalent simultaneous offers (MESO)
value engineering

This chapter addresses the challenges faced by the landscape industry during a recessionary economy. Under economic constraints, sustaining a business requires a multifaceted strategy. The key to success is survival, which means "reaching a sustainable balance between revenue and expenses."[1] "The companies who survive and have financial reserves will weather the storm while retaining most of their personnel and selectively recruiting for future growth."[2] In every instance, the captain of the ship needs to reduce cargo (costs) to guide his/her ship through the storm.

ECONOMIC INDICATORS

The recession that occurred from December 2007 to June 2009 had been the worst in over forty years. Unemployment went from 5.1% in 2007 to 9.5% in June 2009. Significant economic growth was not evident until 2011. It was broader in terms of affecting several economic sectors, and because it involved the banking industry's subprime mortgage crisis, it had a worldwide impact. We can point our finger to high-risk loans and the collapse of the credit market with this recession period, but what do all recessions have in common? Decreased consumer spending, unemployment, foreclosures, losses in investment portfolios, decreases in commercial and residential construction, and limited bank credit.

ECONOMIC CHALLENGES

Surveys of landscape companies indicated the following challenges during a recessionary economy:

- Clients are looking for more for less
- Greater number of companies competing for jobs
- Enhancement budgets reduced
- Maintaining quality and margins

> "Current prices were 30–40% off of asking price."[3] *Josh Denison, Vice President, Denison Landscaping, Waldorf, Maryland, referring to his commercial construction division.*
>
> "Biggest obstacle: Adapting to our client's imperatives to reduce costs and provide the highest level of value for every dollar they spend, providing the best quality at the lowest price, engaging our teams to focus on efficiency and cost savings."[4]

Economic sustainability

The hardest hit segment was design/build, both residential and commercial. Landscape residential and commercial installations are generally scaled down and are significantly less in number due to decreases in construction budgets and home owners' disposable income. A residential design/build company that under normal circumstance would have a six- to eight-month backlog of jobs would have a month if they were lucky. As an enticement to residential clients, a national award-winning company changed their tag line from "Quality Landscapes for Those Who Appreciate It" to "Quality Landscapes at an Affordable Price."

Maintenance contracts on a commercial scale were generally stable. However, landscape maintenance companies usually derive up to $1 additional enhancement expenditure for every dollar of the contract. That income was significantly reduced during the recessionary period.

PROACTIVE STRATEGIES

Instead of taking the ostrich approach of sticking one's head in the sand and waiting for an economic recovery, the proactive companies viewed the negative business environment as an opportunity to add ballast to their ships to keep them afloat. These were the recommendations of some industry leaders and business consultants:

> Companies must become leaner and keep a closer eye on waste and expenses. Those companies who survive during a down economy will be positioned to reap the benefits when the economy turns around. Now is the time to work on retaining clients and building stronger business relationships.[5]
>
> Competition and clients are impacting pricing, resulting in lower prices and tighter profit margins. Address the challenge by lowering indirect and overhead costs in relation to revenues and direct labor. The objective is to provide more value for less money. Work with lower gross margins, higher revenues and stable overhead costs.[6]

Kehoe's last statement had become the mantra of the companies who were sustaining in the recessionary environment. They had addressed and implemented the following cost savings:

- Engaged production teams to focus on efficiency and cost savings
- Extended turnover rate on equipment
- Furloughed employees, e.g., one day per month
- Limited overtime hours, e.g., normally 65/wk down to 43–47 for maintenance
- Reduced overproduction
- Reduced downtime
- Reduced inventory
- Reduced defects
- Monitored accounts receivable

Economic sustainability

- Froze bonuses and raises
- Reduced employee relations activities, e.g., parties, retreats
- Reduced sponsorships
- Restructured personnel, e.g., one mechanic vs. two, two instead of four administrative assistants
- Increased sales volume for account managers
- Negotiated with vendors for lower pricing

Previous chapters have addressed most of the listed cost reductions as a means of increasing a company's gross margin and net profit. There is even more emphasis on cost reduction when revenues are down. It is readily evident that lean management plays a vital role in economic sustainability. Increasing production efficiency and reducing financial waste is the end result of its implementation and subsequent cost savings.

Furloughs, or time off from work without pay, are common in both the public and private sectors. Due to reduced revenues and operating budgets, companies and institutions implement furlough policies. The policies establish the amount of days that an employee would not work within a specific period, e.g., monthly, annually.

MARKETING STRATEGIES

In addition to reducing costs, companies must address increasing revenues. Marketing strategies are the keys to opening up new revenue doors. Some of the strategies that landscape companies have implemented include the following:

- Regional branding
- Emphasizing company's attributes
    - Quality and reputation
    - Certifications
    - Relationships
    - Communication
    - Professionalism
    - Technology
    - Drones
- Entering new market niches
    - Artificial turf fields
    - HOAs (homeowner associations)

Economic sustainability

- Green roofs
- Bioretention facilities
- Biowalls
- Sustainable landscape management
  - Integrated pest management
  - Irrigation management
    - Low-volume sprinklers
    - Drip irrigation
    - Moisture sensors
    - Water audits
  - Propane and electric maintenance equipment
- Outdoor lighting
- Parking lot striping
- Holiday decorating
- Educating clients
  - Sell on value to actual cost
  - Value engineering on every contract
- Engaging the company with the local community through service projects and networking
- Increasing sale efforts
- Monitoring sales leads and having frequent direct contact, quick turnaround, and follow-up
- Networking with community and professional associations
- Increasing marketing budgets
  - Billboards
  - Newsletters
  - Internet
    - Web page
    - Social media
  - Customer incentives

The marketing strategies listed are intended to get companies greater exposure in their existing market through community involvement, with a targeted sales approach directed towards expanding market shares and **branding** to establish a greater identity. Some of these objectives can be fulfilled by advertising, networking with professional and community organizations, maintaining clean vehicles with distinct logos, and having neatly uniformed employees. In a price-sensitive market, it is also important to market a company's professionalism, reputation, quality services, insurance coverage and certifications among employees, and environmental

Economic sustainability

management practices. An adjunct to quality services is **value engineering**, which increases the value of a contracted service by either improving the delivered service (more for less) or reducing its cost.

Environmental services and system construction has been a growth area for landscape companies. If they do not have the expertise in-house, they retain subcontractors who specialize in the installation of such systems, e.g., biowalls for interiorscapes, landscape retaining walls, and storm water management systems, such as bioretention facilities and green roofs (Figures 14.1a–14.1d). Maintenance companies have also incorporated sustainable landscape practices through the use of organic fertilization programs, integrated pest management programs, and irrigation management.

Another approach to marketing is establishing a service menu that can be adapted to a client's budget (Figure 14.1). Referred to as **MESO, multiple equivalent simultaneous offers** enable clients to select options for landscape maintenance services. These range from reducing service frequency, e.g., mowing cuts, to eliminating services, e.g., winter pruning or providing them on an as-needed basis, e.g., weed control. This service option plan is developed through conversations with the client that involve a discussion

Figure 14.1a Bioretention Facilities
Source: Environmental Quality Resources, LLC, Millersville, MD

Figure 14.1b Biowalls, Furbishco, Baltimore, MD

Figure 14.1c Green Roofs, Furbishco, Baltimore, MD

Economic sustainability

Figure 14.1d Retaining Walls, Furbishco, Baltimore, MD

of expectations relative to the site's landscape components. A summary of this marketing approach is as follows:

Prepackaged options (like car models, a pricing menu) based on:

- Conversations with the client
- Relating knowledge of the site
- Maintaining maintenance standards
- Meeting the client's budget

What shouldn't we present?

- Unclear expectations
- A service list without a cost estimate
- Elimination of essential service schedules

Economic sustainability

Table 14.1 Sample MESO (Multiple Equivalent Simultaneous Offers)

Description	Current Scope Quantity	Alternative Scope Quantity
Turf care		
Mowing & trimming	26	24
Edge curbs & walks	13	10
Turf application – early spring	1	1
Turf application – late spring	1	1
Turf application – fall	1	1
Turf application – late fall	1	Delete
Aerate turf	30% of Property	Delete
Overseed turf	30% of Property	Delete
Lime	30% of Property	Delete
Tree & shrub care		
Shrub pruning (no less than 3)	As necessary	3
Winter pruning	Delete	Delete
Tree pruning	1	1
Pre-emergent	1	1
Weeding – hand	As necessary	As necessary
Weed control – beds (spray)	As necessary	As necessary
Weed control cracks & curb (spray)	As necessary	As necessary
Fertilize shrubs	1	Delete
Care of plant material – pest/disease control	As necessary	Monitor only
Spring clean-up	1	1
Edge beds & tree rings	1	1
Spring mulch (dyed mulch)	Dyed Mulch	Regular Mulch
Leaf removal	3	3
Perennial care	As necessary	1

SUMMARY

A recessionary economy impacts all sectors of the landscape industry. Decreases in residential and commercial construction, unemployment, reduced consumer spending, and bank lending restrictions collectively reduce the revenue stream for landscape companies; the hardest hit sector of which is commercial landscape construction and residential design/build. The economic impact on maintenance companies is manifested in reductions of enhancements and client demands for services at a lower price.

Economic sustainability

What is the impact of economic constraints?

Landscape companies are faced with greater competition in the market and clients demanding lower prices for their services. The challenge facing the industry is how to deliver the same quality services at a lower price while maintaining margins. Financial sustainability is jeopardized as reduced revenue is insufficient for meeting existing overhead expenses.

How do companies attain financial sustainability?

Companies must first address production and overhead costs and implement reductions to align with revenues. These reductions start with increasing operation efficiency and reducing downtime and wastes. Overhead budget lines are assessed and reductions applied wherever possible. These would include restructuring personnel in support divisions, such as maintenance and administrative, eliminating salary increases and bonuses, deferring employee relation events or scaling them down, and eliminating or reducing attendance at national conferences. Furloughs, or time off without pay, enable companies to retain personnel. In addition, the elimination or reduction of overtime adds significantly to the reduction of direct costs.

Kevin Kehoe recommends that companies "work towards increasing revenues, lowering margins and maintaining stable overhead costs."[7]

What are the marketing strategies for a negative economy?

Companies need to strengthen their existing relationships and acquire greater market shares by building on their reputation and services. They need to concentrate on branding within their markets to establish a strong identity for their services. Sales proposals need to emphasize value engineering and customized services that provide clients with options, e.g., multiple equivalent simultaneous offers. This marketing approach can be adapted to either maintenance or landscape construction. The intention is to accommodate current client budgets while still meeting the client's expectations.

New market opportunities should be pursued. Design/build and commercial construction companies have concentrated on establishing or expanding maintenance divisions. Green roofs, bioretention systems, athletic turf installation, biowalls, water-conserving irrigation systems, parking lot striping, and holiday decorating are market niches that have increased revenue streams.

Companies are also incorporating new technology, such as drones. Think of their application in the real estate industry. Buyers and realtors can now give video tours of properties

Economic sustainability

Figure 14.2 Phantom-Drone Diagnostics
Source: Bartlett Tree Experts

online. Bartlett Tree Experts utilize drone technology (Figure 14.2) to diagnose tree decline. Landscape companies can utilize drones for monitoring construction projects as well as assessing the health status of large property plant inventory and site topography.

It is essential that companies maintain a high profile through advertising and networking within community and professional associations. Community service projects also contribute to a company's profile by projecting their image as a member of the community.

KNOWLEDGE APPLICATION

1 What marketing strategies have you observed among businesses in your area in response to the recessionary economy?
2 List cost reductions that you could implement in the delivery of quality landscape maintenance services.
3 As a production manager, how would you attain efficiency and cost reduction goals?
4 What companies do you feel have strong branding? How do they achieve it?

Economic sustainability

NOTES

1. Landon Reeve, CEO, Chapel Valley Landscape, Woodbine, MD.
2. Ibid.
3. Josh Denison, Vice President, Denison Landscaping, Waldorf, MD.
4. Scott Brickman, The Brickman Group (currently BrightView).
5. Landon Reeve, CEO, Chapel Valley Landscape Company.
6. Kevin Kehoe, Business Consultant.
7. Ibid.

Appendices

Appendix
A **Landscape-Related Trade Associations**
B **References**
   Business Management
   Leadership
C **Design Build Financial Statements**
   Exterior Installation Financial Statements
   Exterior Maintenance Financial Statements

Source; 2012 Operating Cost Survey, National Association of Landscape Professionals, Herndon, VA.

APPENDIX A

LANDSCAPE-RELATED TRADE ASSOCIATIONS

American Hort
   www.americahort.org
American Society of Landscape Architects
   www.asla.org
International Society of Arboriculture
   www.isa-arbor.com
The Irrigation Association
   www.irrigation.org
Professional Grounds Management Society
   www.pgms.org
National Association of Landscape Professionals
   www.landscapeprofessionals.org
Sports Turf Managers Association
   www.stma.org
Tree Care Industry Association
   www.tcia.org

APPENDIX B

REFERENCES

Business management

Built to Last: Successful Habits of Visionary Companies. James Collins and Jerry Porras, 1994. Cambridge, MA: Harper Business.
Good to Great. Jim Collins, 1958. Cambridge, MA: Harper Business.
Great by Choice. Jim Collins and Morton T. Hansen, 2011. NY: Harper Collins.
How to Price Landscape and Irrigation Projects. James Huston, 1994. Englewood, CO: Smith Huston, Inc., www.jrhuston.biz.
In Search of Excellence: Lessons Learned From America's Best-Run Companies. Thomas J. Peters and Robert Waterman, 1982. NY: Harper & Row.
Landscape Estimating and Contract Administration. Steven Angley, Edward Horsey, and David Roberts, 2002. Albany, NY: Delmar Publishing.
Re-Imagine. Thomas Peters, 2003. Self-published.
The Toyota Way. Jeffrey Like, 2004. NY: McGraw-Hill.
Who Moved My Cheese? Spencer Johnson, 1998. G.P. Putnam's Sons.

Leadership

Contented Cows Give Better Milk. Bill Catlette and Richard Hadden, 2001, Hoboken, NJ: John Wiley and Sons.
Fail Safe Leadership. Linda L. Martin and Dr. David G. Mutchler, 2001. FL: Delta Books.
First, Break All the Rules. Marcus Buckingham and Curt Coffman, 1999. NY: Simon & Schuster.
Leadership and the One-Minute Manager. Ken Blanchard, Patricia Zigarmi, and Drea Zigarmi, 1985. NY: Harper Collins.

APPENDIX C

Design Build Financial Statements
Exterior Installation Financial Statements
Exterior Maintenance Financial Statements

Design/Build

Detailed Results

	Typical Contractor	High-Profit Contractor	Design/ Build Focus	High-Profit Design/ Build	Sales Under $1 Million	Sales $1 - $4 Million	Sales Over $4 Million
Typical Sales $ Volume	2,081,091	2,273,690	1,750,480	2,965,245	875,670	1,704,051	6,092,883
Income Statement (% of sales)							
Direct Costs							
Direct Labor	25.1	24.5	23.6	22.1	23.1	24.5	22.9
Subcontract Labor	6.5	5.3	8.6	12.2	7.2	6.8	11.6
Plant Material	5.9	4.7	7.8	5.9	5.1	8.6	8.0
Hard Material	10.5	8.9	14.5	14.8	14.6	14.1	13.2
All Other Direct Costs	0.8	0.9	0.6	0.8	0.4	0.6	0.7
Total Direct Costs	48.8	44.3	55.1	55.8	50.4	54.6	56.4
Gross Margin	51.2	55.7	44.9	44.2	49.6	45.4	43.6
Equipment Expenses							
Mechanic Salary, Wage & Bonus	0.8	0.9	0.9	0.9	0.7	0.9	1.2
Equipment Rental/Lease	0.5	0.3	0.5	0.1	0.5	0.4	0.6
Fuel & Oil	4.5	4.3	3.6	3.1	3.7	3.8	3.7
Depreciation	3.4	2.6	3.3	2.1	2.4	3.7	2.9
Vehicle/Equipment Insurance	0.8	0.9	0.6	0.5	1.3	0.7	0.4
Repair Parts & Shop Supplies	2.6	2.2	2.7	1.1	3.0	2.2	3.1
Small Equipment	0.1	0.1	0.1	0.4	0.1	0.2	0.1
All Other Equipment Expenses	0.3	0.4	0.2	0.7	0.0	0.3	0.1
Total Equipment Expenses	13.0	11.7	11.9	8.9	11.7	12.2	12.1
Indirect Expenses							
Indirect Labor Wage & Bonus	3.8	2.0	3.9	1.6	3.4	3.8	3.5
Garments/Uniforms	0.2	0.2	0.2	0.1	0.1	0.2	0.0
Tools & Supplies	0.5	0.3	0.5	0.4	0.5	0.4	0.5
Safety & Training	0.2	0.1	0.2	0.1	0.1	0.2	0.1
All Other Indirect Expenses	0.1	0.0	0.1	0.6	0.2	0.2	0.3
Total Indirect Expenses	4.8	2.6	4.9	2.8	4.3	4.8	4.4
G&A Payroll Expenses							
Owners/Officers Salary & Bonus	5.5	6.0	5.3	5.3	8.6	5.1	4.5
Sales Salary, Wage & Commission	4.1	2.9	3.6	3.3	3.1	3.3	4.8
All Other Salary, Wage & Bonus	3.9	5.9	2.6	3.9	3.8	3.9	1.2
Total G&A Salary, Wage & Bonus	13.5	14.8	11.5	12.5	15.5	12.3	10.5
Payroll Taxes (FICA, unemp. & workers' comp.)	4.9	5.2	4.9	4.4	4.9	3.9	4.9
Group Insurance (medical, hospitalization)	1.5	1.5	1.3	0.7	1.5	1.3	1.7
Employee Benefits (pension, retirement)	0.3	0.3	0.4	0.1	0.0	0.5	0.5
Total G&A Payroll Expenses	20.2	21.8	18.1	17.7	21.9	18.0	17.6
Other G&A Expenses							
Rent or Real Estate Ownership	3.2	3.0	2.7	2.3	3.4	3.3	1.7
Communication	0.9	0.8	0.6	0.6	0.9	0.8	0.6
Advertising & Promotion	1.2	1.1	1.2	1.2	1.5	0.9	0.8
Travel & Entertainment	0.3	0.3	0.2	0.2	0.3	0.3	0.3
Business Insurance (liability, casualty)	0.9	0.9	0.7	0.9	0.9	0.9	0.7
Office	0.9	1.0	0.6	0.8	0.7	0.5	0.6
Bad Debt Loss	0.1	0.1	0.0	0.0	0.0	0.0	0.1
All Other G&A Expenses	3.1	3.1	2.4	1.0	2.0	2.3	2.9
Total Other G&A Expenses	10.6	10.3	8.4	7.0	9.7	9.0	7.7
Total Operating Expenses	48.6	46.4	43.3	36.4	47.6	44.0	41.8
Operating Profit	2.6	9.3	1.6	7.8	2.0	1.4	1.8
Other Income	0.1	0.1	0.2	0.1	0.0	0.1	0.2
Interest Expense	0.4	0.2	0.3	0.1	0.5	0.1	0.4
Other Non-Operating Expenses	0.0	0.0	0.0	0.0	0.0	0.0	0.0
Profit Before Taxes	2.3	9.2	1.5	7.8	1.5	1.4	1.6

Design/Build

Detailed Results

	Typical Contractor	High-Profit Contractor	Design/ Build Focus	High-Profit Design/ Build	Sales Under $1 Million	Sales $1 - $4 Million	Sales Over $4 Million
Strategic Profit Model Ratios							
Profit Margin (%)	2.3	9.2	1.5	7.8	1.5	1.4	1.6
Asset Turnover	4.1	5.3	4.1	4.7	2.8	5.5	3.4
Return On Assets (pre-tax %)	9.4	48.8	6.1	36.7	4.2	7.7	5.4
Financial Leverage	2.5	1.5	2.6	1.4	N/A	N/A	N/A
Return On Net Worth (pre-tax %)	23.5	73.2	15.9	51.4	N/A	N/A	N/A
Typical Total $ Assets	507,583	428,998	426,946	630,903	312,739	309,827	1,792,024
Balance Sheet (% of assets)							
Assets							
Cash & Marketable Securities	12.6	32.0	21.0	51.7	25.4	35.0	12.3
Accounts Receivable	29.1	31.1	30.0	31.7	15.3	30.8	30.2
Inventory	2.8	1.1	6.6	3.2	2.1	4.4	15.9
Other Current Assets	4.5	6.9	3.9	2.0	0.5	3.7	6.0
Total Current Assets	49.0	71.1	61.5	88.6	43.3	73.9	64.4
Fixed & Noncurrent Assets	51.0	28.9	38.5	11.4	56.7	20.1	35.6
Total Assets	100.0	100.0	100.0	100.0	100.0	100.0	100.0
Liabilities and Net Worth							
Accounts Payable	12.7	7.9	16.7	6.9	N/A	N/A	N/A
Notes Payable	11.0	6.2	11.2	2.1	N/A	N/A	N/A
Other Current Liabilities	12.1	8.1	13.1	4.4	N/A	N/A	N/A
Total Current Liabilities	35.8	22.2	41.0	13.4	N/A	N/A	N/A
Long-Term Liabilities	24.5	12.6	20.5	12.6	N/A	N/A	N/A
Net Worth or Owner Equity	39.7	65.2	38.5	74.0	N/A	N/A	N/A
Total Liabilities & Net Worth	100.0	100.0	100.0	100.0	100.0	100.0	100.0
Financial Ratios							
Current Ratio	1.6	2.6	1.6	5.6	N/A	N/A	N/A
Quick Ratio	1.2	2.5	1.1	4.7	N/A	N/A	N/A
Accounts Payable To Inventory (%)	126.8	100.0	128.4	92.7	N/A	N/A	N/A
Accounts Payable Payout Period (days)	50.3	34.6	43.6	24.7	N/A	N/A	N/A
Debt To Equity	0.7	0.3	0.7	0.4	N/A	N/A	N/A
EBITTA (% of assets)	11.8	44.4	7.7	25.7	6.8	16.9	7.7
Times Interest Earned	5.0	44.9	3.8	33.2	4.3	4.7	3.4
Asset Productivity							
Inventory Turnover	12.1	16.7	12.0	22.5	N/A	12.9	5.7
Inventory Holding Period (days)	30.1	21.8	30.4	16.8	N/A	28.2	65.0
Sales to Inventory Ratio	75.9	110.3	60.6	110.4	98.4	66.4	24.4
Sales to Fixed Assets	9.5	20.3	12.5	50.3	8.3	20.4	12.3
Collections							
Average Collection Period (days)	26.9	24.4	24.8	20.2	16.0	21.5	36.1
Customer Payment Terms (% of sales)							
Cash & Checks	0.0	0.0	0.0	0.0	0.0	0.0	0.0
Bank Credit Cards	5.0	2.8	5.0	9.4	5.0	5.0	8.5
In-House Credit	95.0	97.2	95.0	90.6	95.0	95.0	91.5
Total Sales	100.0	100.0	100.0	100.0	100.0	100.0	100.0
Cash Flow Cycle (days)							
Average Collection Period (days)	26.9	24.4	24.8	20.2	16.0	21.5	36.1
Plus Inventory Holding Period (days)	30.1	21.8	30.4	16.8	N/A	28.2	65.0
Gross Cash Flow (days)	57.0	46.2	55.2	37.0	N/A	49.7	101.1
Minus A/P Payout Period (days)	50.3	34.6	43.6	24.7	N/A	N/A	N/A
Cash Cycle (days)	6.7	11.6	11.6	12.3	N/A	N/A	N/A
Cash Sufficiency							
Cash to Current Liabilities	24.3	129.4	34.5	312.7	N/A	N/A	N/A
Defensive Interval (days)	16.1	50.2	23.1	117.4	27.0	33.3	20.5
Sales to Working Capital	8.3	6.9	9.3	6.6	N/A	N/A	N/A

Design/Build

Detailed Results

	Typical Contractor	High-Profit Contractor	Design/ Build Focus	High-Profit Design/ Build	Sales Under $1 Million	Sales $1 - $4 Million	Sales Over $4 Million
Customers							
Number of Clients (annual)	300	333	300	305	97	258	706
Sales $ per Landscape Project	7,500	5,956	9,612	7,510	9,612	8,901	17,500
Source of Revenue (% of sales)							
Design/Build	28.0	11.7	75.7	77.5	75.3	77.1	67.0
Exterior Installation	9.9	9.4	0.0	0.0	0.0	0.0	1.8
Exterior Maintenance	38.6	60.3	18.6	19.1	18.8	15.0	13.0
Lawn Care	8.8	9.2	4.0	2.1	0.0	7.9	5.0
Interior Installation & Maintenance	0.0	0.0	0.0	0.0	0.0	0.0	0.0
Irrigation Installation & Maintenance	3.5	7.3	1.1	0.0	3.3	0.0	3.2
Other Revenue	11.2	2.1	0.6	1.3	2.6	0.0	10.0
Total Sales	100.0	100.0	100.0	100.0	100.0	100.0	100.0
Customer Sales (% of sales)							
Residential	68.0	80.0	85.0	95.0	95.0	87.0	70.0
Commercial	32.0	20.0	15.0	5.0	5.0	13.0	30.0
All Other Customers	0.0	0.0	0.0	0.0	0.0	0.0	0.0
Total Sales	100.0	100.0	100.0	100.0	100.0	100.0	100.0
Employee Productivity							
Sales $ per Employee	70,019	73,828	82,860	96,635	75,448	76,476	88,734
Gross Profit $ per Employee	35,464	40,105	36,317	39,905	37,858	35,876	39,816
Payroll $ per Employee	36,605	36,622	39,034	41,071	42,000	38,481	40,913
Personnel Expenses (% of sales)							
Direct Labor	25.1	24.5	23.6	22.1	23.1	24.5	22.9
G&A Payroll Expenses	20.2	21.8	18.1	17.7	21.9	18.0	17.6
Total Personnel Expenses	49.9	49.2	46.5	42.3	49.1	47.2	45.2
FTE Employees at Peak Season							
Owners/Officers	2.0	1.5	1.8	1.9	1.0	1.5	2.5
Sales (incl. sales managers and estimators)	1.8	1.0	1.0	1.4	0.5	1.0	4.5
Landscape Architect/Designers	1.0	0.0	1.5	2.0	0.5	2.0	4.0
Installation/Construction	22.2	25.8	20.9	23.7	7.0	19.5	43.0
Clerical & Administrative	2.0	2.0	1.3	2.3	1.0	1.0	4.0
Total FTE Employees	29.0	30.3	26.5	31.3	10.0	25.0	58.0

Exterior Installation

Detailed Results

	Typical Contractor	High-Profit Contractor	Exterior Installation Focus	High-Profit Exterior Installation	Sales Under $2.5 Million	Sales Over $2.5 Million
Typical Sales $ Volume	2,081,091	2,273,690	2,631,315	2,273,690	938,198	4,210,751
Income Statement (% of sales)						
Direct Costs						
Direct Labor	25.1	24.5	23.9	24.2	23.0	24.9
Subcontract Labor	6.5	5.3	7.3	5.2	7.1	7.4
Plant Material	5.9	4.7	6.2	6.1	6.4	6.1
Hard Material	10.5	8.9	13.3	10.6	10.6	14.7
All Other Direct Costs	0.8	0.9	0.6	0.6	0.4	1.0
Total Direct Costs	**48.8**	**44.3**	**51.3**	**46.7**	**47.5**	**54.1**
Gross Margin	**51.2**	**55.7**	**48.7**	**53.3**	**52.5**	**45.9**
Equipment Expenses						
Mechanic Salary, Wage & Bonus	0.8	0.9	0.9	1.6	1.3	0.0
Equipment Rental/Lease	0.6	0.3	0.3	0.3	0.2	0.7
Fuel & Oil	4.5	4.3	4.0	3.6	3.9	4.0
Depreciation	3.4	2.6	3.5	4.3	3.9	3.2
Vehicle/Equipment Insurance	0.8	0.9	0.7	0.5	0.6	0.7
Repair Parts & Shop Supplies	2.6	2.2	2.8	2.8	3.5	1.8
Small Equipment	0.1	0.1	0.0	0.1	0.0	0.2
All Other Equipment Expenses	0.3	0.4	1.2	0.4	0.8	0.7
Total Equipment Expenses	**13.0**	**11.7**	**13.4**	**13.6**	**14.2**	**11.3**
Indirect Expenses						
Indirect Labor Wage & Bonus	3.8	2.0	3.3	2.9	2.8	4.0
Garments/Uniforms	0.2	0.2	0.2	0.1	0.1	0.2
Tools & Supplies	0.5	0.3	0.7	0.6	1.0	0.7
Safety & Training	0.2	0.1	0.1	0.1	0.1	0.2
All Other Indirect Expenses	0.1	0.0	0.2	0.2	0.1	0.5
Total Indirect Expenses	**4.8**	**2.6**	**4.5**	**3.9**	**4.1**	**5.6**
G&A Payroll Expenses						
Owners/Officers Salary & Bonus	5.5	6.0	3.3	3.7	5.7	N/A
Sales Salary, Wage & Commission	4.1	2.9	3.4	3.8	4.8	N/A
All Other Salary, Wage & Bonus	3.9	5.9	5.1	6.2	3.1	N/A
Total G&A Salary, Wage & Bonus	13.5	14.8	11.8	13.7	13.6	8.8
Payroll Taxes (FICA, unemp. & workers' comp.)	4.9	5.2	4.1	4.5	4.0	6.9
Group Insurance (medical, hospitalization)	1.5	1.5	1.6	1.1	0.8	3.1
Employee Benefits (pension, retirement)	0.3	0.3	0.2	0.2	0.2	0.4
Total G&A Payroll Expenses	**20.2**	**21.8**	**17.7**	**19.5**	**18.6**	**19.2**
Other G&A Expenses						
Rent or Real Estate Ownership	3.2	3.0	3.1	3.8	3.3	2.7
Communication	0.9	0.8	0.7	1.0	0.8	0.7
Advertising & Promotion	1.2	1.1	0.7	1.0	1.4	0.4
Travel & Entertainment	0.3	0.3	0.3	0.5	0.6	0.2
Business Insurance (liability, casualty)	0.9	0.9	0.9	1.1	1.2	0.7
Office	0.9	1.0	0.9	0.8	0.8	0.9
Bad Debt Loss	0.1	0.1	0.0	0.2	0.1	0.0
All Other G&A Expenses	3.1	3.1	2.9	2.9	2.5	3.6
Total Other G&A Expenses	**10.6**	**10.3**	**9.5**	**11.3**	**10.7**	**9.2**
Total Operating Expenses	**48.6**	**46.4**	**45.1**	**48.3**	**47.6**	**45.3**
Operating Profit	**2.6**	**9.3**	**3.6**	**5.0**	**4.9**	**0.6**
Other Income	0.1	0.1	0.0	0.0	0.0	0.1
Interest Expense	0.4	0.2	0.7	1.0	1.1	0.4
Other Non-Operating Expenses	0.0	0.0	0.0	0.0	0.0	0.0
Profit Before Taxes	**2.3**	**9.2**	**2.9**	**4.0**	**3.8**	**0.3**

Exterior Installation

Detailed Results

	Typical Contractor	High-Profit Contractor	Exterior Installation Focus	High-Profit Exterior Installation	Sales Under $2.5 Million	Sales Over $2.5 Million
Strategic Profit Model Ratios						
Profit Margin (%)	2.3	9.2	2.9	4.0	3.8	0.3
Asset Turnover	4.1	5.3	3.6	5.0	5.1	3.0
Return On Assets (pre-tax %)	9.4	48.8	10.4	20.0	19.4	0.9
Financial Leverage	2.5	1.5	3.1	7.0	8.6	2.7
Return On Net Worth (pre-tax %)	23.5	73.2	32.2	140.0	166.8	2.4
Typical Total $ Assets	507,583	428,998	730,921	454,738	183,960	1,403,584
Balance Sheet (% of assets)						
Assets						
Cash & Marketable Securities	12.6	32.0	5.3	18.0	8.1	3.4
Accounts Receivable	29.1	31.1	36.3	35.5	23.6	45.9
Inventory	2.8	1.1	3.7	9.6	14.5	0.0
Other Current Assets	4.5	6.9	4.4	5.0	11.2	2.0
Total Current Assets	49.0	71.1	49.7	68.1	57.4	51.3
Fixed & Noncurrent Assets	51.0	28.9	50.3	31.9	42.6	48.7
Total Assets	100.0	100.0	100.0	100.0	100.0	100.0
Liabilities and Net Worth						
Accounts Payable	12.7	7.9	19.7	19.6	16.8	25.5
Notes Payable	11.0	6.2	13.8	13.4	14.6	10.4
Other Current Liabilities	12.1	8.1	6.3	15.4	18.1	6.6
Total Current Liabilities	35.8	22.2	39.8	48.4	49.5	42.5
Long-Term Liabilities	24.5	12.6	28.2	37.4	38.9	20.7
Net Worth or Owner Equity	39.7	65.2	32.0	14.2	11.6	36.8
Total Liabilities & Net Worth	100.0	100.0	100.0	100.0	100.0	100.0
Financial Ratios						
Current Ratio	1.6	2.6	1.2	1.4	0.9	1.2
Quick Ratio	1.2	2.5	1.0	0.7	0.5	1.1
Accounts Payable To Inventory (%)	126.8	100.0	65.4	49.9	49.9	N/A
Accounts Payable Payout Period (days)	50.3	34.6	49.0	68.3	30.7	72.4
Debt To Equity	0.7	0.3	1.8	5.3	5.7	1.7
EBITTA (% of assets)	11.8	44.4	17.6	30.5	25.3	-1.1
Times Interest Earned	5.0	44.9	6.0	7.5	6.7	N/A
Asset Productivity						
Inventory Turnover	12.1	16.7	14.4	10.6	3.4	N/A
Inventory Holding Period (days)	30.1	21.8	25.3	64.0	107.4	N/A
Sales to Inventory Ratio	75.9	110.3	67.7	58.2	21.6	N/A
Sales to Fixed Assets	9.5	20.3	8.3	10.9	8.2	8.3
Collections						
Average Collection Period (days)	26.9	24.4	34.8	30.2	12.7	53.3
Customer Payment Terms (% of sales)						
Cash & Checks	0.0	0.0	0.0	0.0	0.0	0.0
Bank Credit Cards	5.0	2.8	1.0	2.8	1.0	0.0
In-House Credit	95.0	97.2	99.0	97.2	99.0	100.0
Total Sales	100.0	100.0	100.0	100.0	100.0	100.0
Cash Flow Cycle (days)						
Average Collection Period (days)	26.9	24.4	34.8	30.2	12.7	53.3
Plus Inventory Holding Period (days)	30.1	21.8	25.3	64.0	107.4	N/A
Gross Cash Flow (days)	57.0	46.2	60.1	94.2	120.1	N/A
Minus A/P Payout Period (days)	50.3	34.6	49.0	68.3	30.7	72.4
Cash Cycle (days)	6.7	11.6	11.1	25.9	89.4	N/A
Cash Sufficiency						
Cash to Current Liabilities	24.3	129.4	10.6	4.8	7.7	15.1
Defensive Interval (days)	16.1	50.2	10.6	3.0	10.4	10.6
Sales to Working Capital	8.3	6.9	20.7	6.0	0.7	29.6

Exterior Installation — Detailed Results

	Typical Contractor	High-Profit Contractor	Exterior Installation Focus	High-Profit Exterior Installation	Sales Under $2.5 Million	Sales Over $2.5 Million
Customers						
Number of Clients (annual)	300	333	192	128	192	199
Sales $ per Landscape Project	7,500	5,956	12,032	8,730	7,500	21,115
Source of Revenue (% of sales)						
Design/Build	28.0	11.7	3.2	10.0	15.5	0.8
Exterior Installation	9.9	9.4	57.2	61.2	55.5	59.8
Exterior Maintenance	38.6	60.3	23.4	23.4	15.0	26.0
Lawn Care	8.8	9.2	3.4	3.0	1.5	3.8
Interior Installation & Maintenance	0.0	0.0	0.0	0.0	0.0	0.0
Irrigation Installation & Maintenance	3.5	7.3	5.0	0.0	3.0	5.3
Other Revenue	11.2	2.1	7.8	2.4	9.5	4.3
Total Sales	100.0	100.0	100.0	100.0	100.0	100.0
Customer Sales (% of sales)						
Residential	68.0	80.0	62.0	70.0	72.5	22.0
Commercial	32.0	20.0	38.0	30.0	27.5	78.0
All Other Customers	0.0	0.0	0.0	0.0	0.0	0.0
Total Sales	100.0	100.0	100.0	100.0	100.0	100.0
Employee Productivity						
Sales $ per Employee	70,019	73,828	78,668	78,668	63,169	85,538
Gross Profit $ per Employee	35,464	40,105	35,357	43,112	32,957	35,357
Payroll $ per Employee	36,605	36,622	33,153	29,467	23,408	40,682
Personnel Expenses (% of sales)						
Direct Labor	25.1	24.5	23.9	24.2	23.0	24.9
G&A Payroll Expenses	20.2	21.8	17.7	19.5	18.6	19.2
Total Personnel Expenses	49.9	49.2	45.8	48.2	45.7	48.1
FTE Employees at Peak Season						
Owners/Officers	2.0	1.5	1.0	1.0	1.0	2.0
Sales (incl. sales managers and estimators)	1.8	1.0	2.0	2.0	0.5	2.0
Landscape Architect/Designers	1.0	0.0	1.0	1.0	1.0	0.0
Installation/Construction	22.2	25.8	23.0	15.0	13.6	46.5
Clerical & Administrative	2.0	2.0	3.5	4.0	2.0	4.0
Total FTE Employees	29.0	30.3	30.5	23.0	18.1	54.5

Exterior Maintenance

Detailed Results

	Typical Contractor	High-Profit Contractor	Exterior Maint. Focus	High-Profit Exterior Maint.	Sales Under $1 Million	Sales $1 - $5 Million	Sales Over $5 Million
Typical Sales $ Volume	2,081,091	2,273,690	2,136,900	2,275,046	755,868	1,928,434	7,184,946
Income Statement (% of sales)							
Direct Costs							
Direct Labor	25.1	24.5	29.6	29.6	30.7	27.5	28.9
Subcontract Labor	6.5	5.3	4.4	2.9	2.8	6.1	5.7
Plant Material	5.9	4.7	5.2	3.6	3.8	4.5	5.2
Hard Material	10.5	8.9	8.2	6.7	9.4	7.1	9.0
All Other Direct Costs	0.8	0.9	0.9	0.9	1.3	0.4	1.2
Total Direct Costs	48.8	44.3	48.3	43.7	48.0	45.6	50.0
Gross Margin	51.2	55.7	51.7	56.3	52.0	54.4	50.0
Equipment Expenses							
Mechanic Salary, Wage & Bonus	0.8	0.9	1.1	1.7	0.0	1.4	1.4
Equipment Rental/Lease	0.5	0.3	0.5	0.4	0.3	0.4	0.8
Fuel & Oil	4.5	4.3	4.9	4.7	4.7	5.3	4.9
Depreciation	3.4	2.6	3.4	3.9	2.7	3.5	3.0
Vehicle/Equipment Insurance	0.8	0.9	0.9	0.9	0.9	1.2	0.8
Repair Parts & Shop Supplies	2.6	2.2	3.0	3.2	3.2	3.4	2.4
Small Equipment	0.1	0.1	0.2	0.0	0.0	0.2	0.1
All Other Equipment Expenses	0.3	0.4	0.2	0.2	0.0	0.3	0.1
Total Equipment Expenses	13.0	11.7	14.2	15.0	11.8	15.7	13.5
Indirect Expenses							
Indirect Labor Wage & Bonus	3.8	2.0	3.3	0.5	0.4	4.0	5.5
Garments/Uniforms	0.2	0.2	0.2	0.2	0.7	0.2	0.3
Tools & Supplies	0.5	0.3	0.4	0.3	1.0	0.7	0.3
Safety & Training	0.2	0.1	0.2	0.2	0.4	0.2	0.1
All Other Indirect Expenses	0.1	0.0	0.0	0.1	0.0	0.0	0.1
Total Indirect Expenses	4.8	2.6	4.1	1.3	2.5	5.1	6.3
G&A Payroll Expenses							
Owners/Officers Salary & Bonus	5.5	6.0	5.8	6.0	7.5	6.7	3.9
Sales Salary, Wage & Commission	4.1	2.9	4.7	2.0	0.0	3.4	4.9
All Other Salary, Wage & Bonus	3.9	5.9	3.6	6.1	0.9	3.9	4.5
Total G&A Salary, Wage & Bonus	13.5	14.8	14.1	14.1	8.4	14.0	13.3
Payroll Taxes (FICA, unemp. & workers' comp.)	4.9	5.2	5.3	6.4	6.5	6.2	4.5
Group Insurance (medical, hospitalization)	1.5	1.5	2.0	1.2	2.7	1.5	1.7
Employee Benefits (pension, retirement)	0.3	0.3	0.2	0.1	0.0	0.1	0.4
Total G&A Payroll Expenses	20.2	21.8	21.6	21.8	17.6	21.8	19.9
Other G&A Expenses							
Rent or Real Estate Ownership	3.2	3.0	3.4	3.1	5.6	3.6	2.8
Communication	0.9	0.8	0.8	0.7	1.4	0.9	0.7
Advertising & Promotion	1.2	1.1	0.7	0.7	1.0	0.8	0.9
Travel & Entertainment	0.3	0.3	0.3	0.2	0.6	0.3	0.2
Business Insurance (liability, casualty)	0.9	0.9	0.7	0.7	1.0	0.7	0.7
Office	0.9	1.0	0.8	0.8	1.2	1.0	0.7
Bad Debt Loss	0.1	0.1	0.1	0.2	0.2	0.0	0.2
All Other G&A Expenses	3.1	3.1	3.1	3.7	5.8	3.7	2.4
Total Other G&A Expenses	10.6	10.3	9.9	10.1	16.8	11.0	8.6
Total Operating Expenses	48.6	46.4	49.8	48.2	48.7	53.6	48.3
Operating Profit	2.6	9.3	1.9	8.1	3.3	0.8	1.7
Other Income	0.1	0.1	0.2	0.3	0.1	0.1	0.3
Interest Expense	0.4	0.2	0.3	0.3	0.1	0.4	0.2
Other Non-Operating Expenses	0.0	0.0	0.0	0.0	0.0	0.0	0.0
Profit Before Taxes	2.3	9.2	1.8	8.1	3.3	0.5	1.8

Exterior Maintenance

Detailed Results

	Typical Contractor	High-Profit Contractor	Exterior Maint. Focus	High-Profit Exterior Maint.	Sales Under $1 Million	Sales $1 - $5 Million	Sales Over $5 Million
Strategic Profit Model Ratios							
Profit Margin (%)	2.3	9.2	1.8	8.1	3.3	0.5	1.8
Asset Turnover	4.1	5.3	4.7	4.7	4.5	4.6	4.8
Return On Assets (pre-tax %)	9.4	48.8	8.5	38.1	14.8	2.3	8.6
Financial Leverage	2.5	1.5	2.3	N/A	N/A	N/A	N/A
Return On Net Worth (pre-tax %)	23.5	73.2	19.5	N/A	N/A	N/A	N/A
Typical Total $ Assets	507,583	428,998	454,660	484,052	167,971	419,225	1,496,864
Balance Sheet (% of assets)							
Assets							
Cash & Marketable Securities	12.6	32.0	13.4	24.6	20.3	10.2	11.8
Accounts Receivable	29.1	31.1	28.9	39.2	18.7	29.8	45.1
Inventory	2.8	1.1	0.0	0.0	0.0	0.2	3.1
Other Current Assets	4.5	6.9	4.2	8.3	1.7	7.5	5.7
Total Current Assets	49.0	71.1	46.5	72.1	40.7	47.7	65.7
Fixed & Noncurrent Assets	51.0	28.9	53.5	27.9	59.3	52.3	34.3
Total Assets	100.0	100.0	100.0	100.0	100.0	100.0	100.0
Liabilities and Net Worth							
Accounts Payable	12.7	7.9	11.9	N/A	N/A	N/A	N/A
Notes Payable	11.0	6.2	9.5	N/A	N/A	N/A	N/A
Other Current Liabilities	12.1	8.1	10.8	N/A	N/A	N/A	N/A
Total Current Liabilities	35.8	22.2	32.2	N/A	N/A	N/A	N/A
Long-Term Liabilities	24.5	12.6	24.5	N/A	N/A	N/A	N/A
Net Worth or Owner Equity	39.7	65.2	43.3	N/A	N/A	N/A	N/A
Total Liabilities & Net Worth	100.0	100.0	100.0	100.0	100.0	100.0	100.0
Financial Ratios							
Current Ratio	1.6	2.6	1.4	N/A	N/A	N/A	N/A
Quick Ratio	1.2	2.5	1.2	N/A	N/A	N/A	N/A
Accounts Payable To Inventory (%)	126.8	100.0	278.9	N/A	N/A	N/A	N/A
Accounts Payable Payout Period (days)	50.3	34.6	78.1	N/A	N/A	N/A	N/A
Debt To Equity	0.7	0.3	0.7	N/A	N/A	N/A	N/A
EBITTA (% of assets)	11.8	44.4	11.8	41.6	21.4	4.4	12.4
Times Interest Earned	5.0	44.9	7.3	20.9	8.3	6.0	6.9
Asset Productivity							
Inventory Turnover	12.1	16.7	16.8	N/A	43.9	28.1	16.5
Inventory Holding Period (days)	30.1	21.8	21.8	N/A	29.2	15.5	22.1
Sales to Inventory Ratio	75.9	110.3	143.7	216.2	442.2	206.3	124.4
Sales to Fixed Assets	9.5	20.3	10.6	17.0	8.9	9.3	13.7
Collections							
Average Collection Period (days)	26.9	24.4	28.0	25.3	23.9	26.2	34.3
Customer Payment Terms (% of sales)							
Cash & Checks	0.0	0.0	0.0	0.0	0.6	0.0	0.0
Bank Credit Cards	5.0	2.8	4.0	1.5	4.0	5.0	4.5
In-House Credit	95.0	97.2	96.0	98.5	95.4	95.0	95.5
Total Sales	100.0	100.0	100.0	100.0	100.0	100.0	100.0
Cash Flow Cycle (days)							
Average Collection Period (days)	26.9	24.4	28.0	25.3	23.9	26.2	34.3
Plus Inventory Holding Period (days)	30.1	21.8	21.8	N/A	29.2	15.5	22.1
Gross Cash Flow (days)	57.0	46.2	49.8	N/A	53.1	41.7	56.4
Minus A/P Payout Period (days)	50.3	34.6	78.1	N/A	N/A	N/A	N/A
Cash Cycle (days)	6.7	11.6	-28.3	N/A	N/A	N/A	N/A
Cash Sufficiency							
Cash to Current Liabilities	24.3	129.4	25.9	N/A	N/A	N/A	N/A
Defensive Interval (days)	16.1	50.2	14.3	36.6	33.7	6.0	11.6
Sales to Working Capital	8.3	6.9	6.0	N/A	N/A	N/A	N/A

Exterior Maintenance

Detailed Results

	Typical Contractor	High-Profit Contractor	Exterior Maint. Focus	High-Profit Exterior Maint.	Sales Under $1 Million	Sales $1 - $5 Million	Sales Over $5 Million
Customers							
Number of Clients (annual)	300	333	297	350	150	250	448
Sales $ per Landscape Project	7,500	5,956	7,500	7,588	2,914	6,750	13,000
Source of Revenue (% of sales)							
Design/Build	28.0	11.7	9.1	0.0	6.7	13.4	6.9
Exterior Installation	9.9	9.4	10.4	20.8	0.0	8.6	14.5
Exterior Maintenance	38.6	60.3	66.3	72.0	85.9	56.8	63.7
Lawn Care	8.8	9.2	5.2	0.0	6.6	4.4	2.6
Interior Installation & Maintenance	0.0	0.0	0.0	0.0	0.0	0.0	0.0
Irrigation Installation & Maintenance	3.5	7.3	4.2	6.4	0.5	5.0	3.1
Other Revenue	11.2	2.1	4.8	0.8	0.3	11.8	9.2
Total Sales	100.0	100.0	100.0	100.0	100.0	100.0	100.0
Customer Sales (% of sales)							
Residential	68.0	80.0	54.6	32.2	82.0	40.0	24.2
Commercial	32.0	20.0	45.4	67.8	18.0	60.0	75.8
All Other Customers	0.0	0.0	0.0	0.0	0.0	0.0	0.0
Total Sales	100.0	100.0	100.0	100.0	100.0	100.0	100.0
Employee Productivity							
Sales $ per Employee	70,019	73,828	62,185	59,674	57,966	55,956	72,053
Gross Profit $ per Employee	35,464	40,105	32,743	37,843	35,476	30,555	37,783
Payroll $ per Employee	36,605	36,622	34,503	36,330	29,042	29,929	37,763
Personnel Expenses (% of sales)							
Direct Labor	25.1	24.5	29.6	29.6	30.7	27.5	28.9
G&A Payroll Expenses	20.2	21.8	21.6	21.8	17.6	21.8	19.9
Total Personnel Expenses	49.9	49.2	55.6	53.6	48.7	54.7	55.7
FTE Employees at Peak Season							
Owners/Officers	2.0	1.5	1.0	1.5	1.0	1.0	2.0
Sales (incl. sales managers and estimators)	1.8	1.0	2.0	1.5	0.4	2.0	4.3
Landscape Architect/Designers	1.0	0.0	0.3	0.0	0.0	0.5	1.0
Installation/Construction	22.2	25.8	26.7	30.2	7.6	26.5	87.7
Clerical & Administrative	2.0	2.0	2.0	1.8	1.0	1.0	5.0
Total FTE Employees	29.0	30.3	32.0	35.0	10.0	31.0	100.0

INDEX

Note: Page numbers in italic indicate a figure and page numbers in bold indicate a table on the corresponding page.

4P Model 203–206, *204*, *206*, 220; people and partners 205; philosophy 203; problem solving 205–206; process 203–205, *204*
5-S process 206–210, *207*, *208*, *209*, *210*; shine 208, 220; sort 207; standardize 207, 208, 210, 220; straighten 207, 220; sustain 207, 208, 210, 220

accountability, project management 104–105
accounting systems: advantages of different types **6**, **7**, **8**; billing basis type 6–7, **7**; cash basis type **5**, 5–6; disadvantages of different types **6**, **7**, **8**; knowledge application exercises 18; percentage of completion basis type 7–8, **8**; profile *4*; software applications 130–137, *131*, *132*, *133*, *134*, *135*, *136*; summary 17–18; value of 2–3
account managers 27
accounts, chart of 9
accounts, coding of **10**, 10–17, **11**, **12**, **13**, **14**, **15**, **16**
accounts payable 5; balance sheet components 76, **77**; payout period 118–119; *see also* balance sheet; liabilities
accounts receivable 76, 78, 110–114, 117–118, 125; balance sheet components 76, **77**; structuring for accountability 5, 7, 8; *see also* assets; average collection period; balance sheet; cash basis type accounting system; cash flow considerations; financial ratios
accrual accounting system: billing basis type 6–7, **7**; percentage of completion basis type 7–8, **8**
accrued taxes and liabilities 76, **77**; *see also* balance sheet
ACPH *see* acquisition cost per hour
acquisition cost per hour 51–52
ADA *see* Americans with Disabilities Act
Americans with Disabilities Act 150
as-builts 100; *see also* project management
asset capital 29–30; *see also* profit-based method of budget development
assets 5, 10–11, **11**, 12–13; current 10, **11**; fixed 10, **11**; *see also* balance sheet
average collection period 117–118
average holding period 117; *see also* inventory turnover ratio

balance sheet **10**, 10–13, **11**, **12**, **13**; components 76–78, **77**, **115**; *see also* financial ratios; income statement
BE *see* breakeven point
best management practices 97, 105–106
billing basis accounting system 6–7, 7
BMP *see* best management practices
branding 247–248; *see also* marketing strategies
breakeven point 40–44, **41**, **42**, **43**
budget: capital 20; cash flow 20; definition 20; development process 28–37, **29**, **31**, **32**, **33**, **35**; forecasting 28–29, **29**; operating 20–28, **22**, **23**, **24**, **25**, **26**, **27**, **28**
budget, development process 28–37, **29**, **31**, **32**, **33**, **35**; growth rate method 28–29; inflation rate method 28; knowledge application exercises 36–37; percentage method 28–29; profit-based method **29**, 29–35, **31**, **32**, **33**, **35**; summary 35–36
building projects: estimating considerations 65–69, **69**; landscape construction estimate example **69**

capital: asset 29–30; budget 20; working 25
cash basis type accounting system **5**, 5–6
cash flow considerations: budget 20; negative type 2, 20, 78, 82, 110, 113, 119; positive type 2, 20, 82; reports 82–83

Index

cash to current liabilities ratio 111–113; *see also* financial ratios
certification programs 233–234; *see also* professional development
certified public accountant 4
chart of accounts 9
chemical systems 196–197
client presentations 192–193
closure, project management 100, 104
communication: multilingual 191; productivity basics 190–192
complement markup **41**, 41–42; *see also* pricing considerations
computerized estimating programs 70; *see also* estimating considerations
contingency clause 65; *see also* estimating considerations
contracts: maintenance 63–65; management of 92–93; retainage fee 93; work in progress schedule 93; *see also* estimating considerations
costs, direct 13–14, 28–29; *see also* forecasting, operating budget; pricing considerations
costs, indirect 15, 28–29; *see also* forecasting, operating budget
CPA *see* certified public accountant
current assets 10, **11**; balance sheet components 76, **77**
current liabilities 11; balance sheet components 76, **77**; *see also* liabilities
current ratio 110, 111–113; *see also* financial ratios

depreciation 30; *see also* profit-based method of budget development
design software 139–140
digitizer 59
direct costs 13–14, 28–29; estimating considerations 59, 66–67, 73; income statement components **79**, 79–81, **80**, **81**; material tracking 86–88; pricing considerations 39, **40**, **41**, **45**, **47**, 51, 53–54; *see also* forecasting, operating budget
DORM *see* dual overhead rate method
dual overhead rate method: dual rate formulas 48–50; materials to labor ratio 47–48;
overhead costs 46–50, **47**, **48**, **49**; overhead weighting factors 47–48, 55
dual rate formulas 48–50

economic challenges 244–245
economic indicators 244
economic proactive strategies 245–246
economic sustainability: challenges 244–245, 252; economic indicators 244; knowledge application exercises 253; marketing strategies 246–251, *248*, *249*, *250*, **251**, 252–253; proactive strategies 245–246; summary 251–253, *253*
Eiji Toyoda 201
employee: evaluations 159–160; recognition techniques 164–166; retention considerations 161–164; terminations 166–172, 173–174; *see also* managing human assets
employee terminations 166–172, 173–174; example employee clauses 167–168; exit interview 168–170
equipment pricing considerations 51–53, **52**
estimating considerations: building projects 65–69, **69**, 194; computerized estimating programs 70; direct costs 59, 66–67, 73; indirect costs 59, 67, 73; job cost management 85–86; key estimating components 58–60; knowledge application exercise 73; landscape construction estimate example **69**; maintenance contracts 63–65; markups 59–60, 67, 70–73; production rates 60–63, **61**, **62**; profit measures 60, 67–68, 72–73; software applications 137–138; summary 72–73; time and materials estimate 70–72, **71**, **72**
execution, project management 99–100, *101*, *102*, *103*
exit interview 168–170
expenses introduction 2–4; *see also* direct costs; overhead costs

final selling price **41**, 41–44, **42**, **43**, 55; *see also* pricing considerations
financial management: balance sheet components 76–78, **77**; cash flow reports 82–83; contracts, management of 92–93;

Index

estimating considerations 85–86; income statement components **79**, 79–81, **80**, **81**; job cost reports 90–92, **91**, **92**; knowledge application exercises 95; material tracking 86–88; production rates 88–90, **89**; summary 94–95

financial ratios: accounts payable payout period 118–119; average collection period 117–118; cash to current liabilities ratio 111–113; current ratio 110, 111–113; debt ratio 113–114; fixed assets 120–121; gross margin 121–122; inventory turnover ratio 114, 116–117; knowledge application exercises 126–127; labor costs 119–120; liquidity ratio 109–110; owner equity 124–125; profit margin 122–123; quick ratio 110, 111–113; return on assets 123; return on net worth 124; summary 125–126; see also balance sheet; income statement

fixed assets 10, **11**; balance sheet components 76, **77**; productivity ratios 120–121

forecasting, operating budget 22–28, **23**, **24**, **25**, **26**, **27**, **28**; account managers 27; budget parameters 28–29, **29**; personnel requirements 26–27, **27**; profit centers **26**, 26–27, **27**; sales and net profit determination 24; sales requirements 25–26, **26**; working capital 25; see also operating budget

FTE see full-time employees

fuel consumption 193–194; see also productivity basics

full-time employees 26

furloughs 246; see also economic sustainability

general and administrative overhead costs **14–15**

geographical information system 143

GIS see geographical information system

global navigation satellite system 142–143; see also mobile resource management

global positioning system 141, 194–195; see also mobile resource management

GNSS see global navigation satellite system

GPS see global positioning system

gross margin 121–122

growth rate method budget process 28–29

H-2B visa 157–158; see also hiring considerations

hiring considerations: foreign labor 157–158; H-2B visa 157–158; interviews 156–157, 158–159; job description components 146–151; legal concerns 156; orientation 159

income statement 13–17, **14**, **15**, **16**; components **79**, 79–81, **80**, **81**, **116**; cost structure 33; gross margin 33; profit-based method of budget process 30, 32, **32**, 33–34; see also balance sheet; financial ratios

indirect costs 15, 28 29; estimating considerations 59, 67, 73; income statement components **79**, 79–81, **80**, **81**; see also forecasting, operating budget

inertial measurement unit 141–142

inflation rate method budget process 28

initiation and scope assessment 98–99; see also project management

internship programs 153–155

interviews, employee 156–157, 158–159; see also hiring considerations

INV see inverse factor

inventory 5, 18; see also balance sheet; overhead costs

inventory turnover ratio: financial ratios 114, 116–117; see also average holding period

inverse factor 42–43, **43**; see also project management

invoice 6–8, **7**, 17; see also billing basis accounting system; liabilities; percentage of completion basis type accounting system

ITR see inventory turnover ratio

job cost management: contracts, management of 92–93; estimating considerations 85–86; job cost reports 90–92, **91**, **92**, *103*; material tracking 86–88; production rates 88–90, **89**

job cost reports 90–92, **91**, **92**, *103*

job description components 146–151; see also managing human assets

JP Horizons Better Results Campaign 229; see also professional development

Index

kaizen 187
kaizen event 210–215, *211, 212, 214, 215,* 221; sample scenario 216–219; *see also* lean manufacturing
K.A.S.H.© system 230–232; *see also* professional development
knowledge application exercises: accounting systems 18; budget, development process 36–37; economic sustainability 253; estimating considerations 73; financial management 95; financial ratios 126–127; leadership 242; lean manufacturing 221; managing human assets 174; pricing considerations 55–56; productivity basics 199; professional development 236; project management 106–107; software applications 144

labor: burden 45–46; costs 119–120; hourly rate selling price 46; markup 45–46; overhead markup 45–46; *see also* income statement; overhead costs; productivity ratios
landscape industry certified: manager 234; technician 233–234; *see also* professional development
leadership: knowledge application exercises 242; mission statements 239–240; multilevel process 240–241; results-oriented process 238–239; summary 241–242; vision statements 239–240
lead time 100; *see also* project management
lean manufacturing: 4P Model 203–206, *204, 206,* 220; 5-S process 206–210, *207, 208, 209, 210*; applicable processes 202–203; kaizen event 210–215, *211, 212, 214, 215,* 221; knowledge application exercises 221; non-value-added time 201, 205–206, 216–217, 220; profit per dollar of sales 216; sales per dollar of salaries 216; sample lean application scenario 216–219; summary 220–221; Toyota Production System 201–202, 220; value-added time 201, 217, 220
ledger 2, *3*
liabilities 11–13, **12, 13**; current 11; long-term 11
LiDAR *see* light detection and ranging

light detection and ranging 141–143
liquidity ratio *see* financial ratios
LOM *see* labor overhead markup
long-term liabilities 11; balance sheet components 76, **77**; *see also* liabilities

maintenance contracts 63–65
management systems: PDCA Cycle 184–186; Six Sigma 183–184; Total Quality Management 182–183
managing human assets: Americans with Disabilities Act 150; employee evaluations 159–160, 172; employee recognition techniques 164–166; employee retention considerations 161–164, 173; employee terminations 166–172, 173–174; example employee clauses 167–168; exit interview 168–170; foreign labor 157–158; internship programs 153–155; interviews 156–157, 158–159; job description components 146–151, 172; knowledge application exercises 174; legal concerns 156, 166–172, 173; millennial-generation employee considerations 163–164; orientation 159; positive company culture 161, 173; recruitment 151–153, 172; summary 172–174; *see also* productivity basics
marketing strategies 246–251, *248, 249, 250,* **251**; branding 247–248; multiple equivalent simultaneous offers 248–251, **251**; value engineering 247; *see also* economic sustainability
markups: estimating considerations 59–60, 67, 70–73; overhead costs 40–44, **41, 42**
material tracking *see* job cost management
MESO *see* multiple equivalent simultaneous offers
millennial-generation employee considerations 163–164
mission statements 239–240; *see also* leadership
mobile mapping *see* mobile resource management
mobile resource management 141–143, *142*; geographical information system 143; global navigation satellite system 141–143; global positioning system 141, 194–195; light detection and ranging 141–143

Index

monitoring and control *see* project management
MORS *see* multiple overhead recovery system
multilevel process 240–241; *see also* leadership
multiple equivalent simultaneous offers 248–251, **251**
multiple overhead recovery system 50

negative cash flow considerations 2, 20, 78, 82, 110, 113, 119
net profit 24; *see also* forecasting, operating budget
networking *see* professional development
net worth **12**, 12–13, **13**
non-value-added time 201, 205–206, 216–217, 220; *see also* lean manufacturing

OHM *see* overhead markup
on-screen takeoffs 59; *see also* takeoffs
on-the-job training *see* training programs
operating budget 20–28, *22*, **23**, **24**, **25**, **26**, **27**, **28**; definition 20–22, *22*; forecasting 22–28, **23**, **24**, **25**, **26**, **27**, **28**
OPPH *see* overhead and profit per hour
overhead and profit per hour 50
overhead costs 3, 6, 9, **14**, 30, **31**, 39; dual overhead rate method 46–50, **47**, **48**, **49**; equipment pricing considerations 51–53, **52**; income statement components **79**, 79–81, **80**, **81**; labor markup 45–46; material markup 40–44, **41**, **42**; multiple overhead recovery system 50; overhead and profit per hour 50; overhead markup 40–42, **41**, **42**; profit-based method of budget development 30, **31**, 32; *see also* income statement
overhead markup 40–42, **41**, **42**
owner equity: balance sheet components 76, **77**; profitability ratios 124–125

PDCA Cycle 184–186
percentage method budget process 28–29
percentage of completion basis type accounting system 7–8, **8**
planning/pre-engineering *see* project management
plant growth regulators **196**
PM *see* profit markup

positive company culture 161
positive cash flow considerations 2, 20, 82
pricing considerations 3, 6, 9, **14**, 30, **31**, 39; breakeven point 40–44, **41**, **42**, **43**; complement markup **41**, 41–42; direct costs 39, **40**, **41**, **45**, **47**, 51, 53–54; dual overhead rate method 46–50, **47**, **48**, **49**; dual rate formulas 48–50; equipment pricing considerations 51–53, **52**; final selling price **41**, 41–44, **42**, **43**, 55; inverse factor 42–43, **43**; knowledge application exercises 55–56; labor markup 45–46; material markup 40–44, **41**, **42**; materials to labor ratio 47–48; overhead costs 46–50, **47**, **48**, **49**; overhead markup 40–42, **41**, **42**; overhead weighting factors 47–48, 55; profit-based method of budget development 30, **31**, 32; profit considerations 39–47, **41**, **43**, **44**, **46**
process flowcharting 186
process management: kaizen 187; productivity standards scorecard 188; quality standards 188
production factors 177–179, **179**, 197
production inefficiencies 180, 197–198
production management: company policies 181; salary, status and security 181–182; supervision 181; theories of motivation 180–181; working conditions 181
production rates: estimating considerations 60–63, **61**, **62**; job cost management 88–90, **89**
productivity basics: chemical systems 195–197, 199; client presentations 192–193; communication 190–192; company policies 181; definition 176–177; estimating 194; fuel consumption 193–194; global positioning system 141, 194–195, 198; kaizen 187; knowledge application exercises 199; PDCA Cycle 184–186; personality test illustration **179**; process flowcharting 186; production factors 177–179, **179**, 197; production inefficiencies 180, 197–198; productivity improvement 186–187; productivity standards scorecard 188; salary, status and security 181–182; supervision 181; theories of motivation 180–181;

272

Index

working conditions 181; *see also* estimating considerations; managing human assets
productivity improvement 186–187
productivity ratios: fixed assets 120–121; labor costs 119–120
productivity standards scorecard 188
professional development: knowledge application exercises 236; landscape industry certified manager 234; landscape industry certified technician 233–234; seminars and workshops 227–229, 235; summary 234–236; trade associations 224–226, 235, 256, 258; training programs 229–232, 235
profit 5–8, 12–15, 17; profit-based method of budget development **29**, 29–35, **31**, **32**, **33**, **35**; *see also* income statement
profitability ratios: gross margin 121–122; owner equity 124–125; profit margin 122–123; return on assets 123; return on net worth 124
profit-based method of budget development **29**, 29–35, **31**, **32**, **33**, **35**; asset capital 29–30; cost structure 33; depreciation 29–30; gross margin 33; income statement 30, 32, **32**, 33–34; monthly revenues and expenses report 34–35, **35**; overhead expense sheet 30, **31**, 32; principle 29; strategic plan 30; watchdogs 30, **31**; working capital 29; zero-based budgeting 30, 33
profit centers 13–15, 18; forecasting, operating budget **26**, 26–27, **27**; full-time employees 26; *see also* job cost management
profit margin 122–123; *see also* profitability ratios
profit markup 40–46, **41**, **42**, **43**; *see also* pricing considerations
profit measures 60, 67–68, 72–73
profit per dollar of sales 216; *see also* lean manufacturing
project management: accountability 104–105; as-builts 100; best management practices 97, 105–106; closure 100, 104; definition 97; execution 99–100, *101*, *102*, *103*; initiation and scope assessment 98–99; knowledge application exercises 106–107; lead time 100; monitoring and control 99; planning/pre-engineering 99; summary 105–106

quality standards 188
quick ratio 110; financial ratios 110, 111–113

recruitment 151–153
results-oriented leadership 238–239
retainage fee 93
retainage monies 8; *see also* percentage of completion basis type accounting system
return on assets 123; *see also* profitability ratios
return on net worth 124; *see also* profitability ratios
revenue considerations: billing basis accounting 7; cash basis accounting 5–6; percentage of completion basis accounting 8; retainage monies 8; *see also* income statement
revenues and expenses report, monthly 34–35, **35**; *see also* profit-based method of budget development
ROA *see* return on assets

salary, status and security 181–182
sales per dollar of salaries 216; *see also* lean manufacturing
sales requirements 25–26, **26**; *see also* forecasting, operating budget
Six Sigma 183–184
smartphone applications 138–139
software applications 141–143, *142*; accounting systems 130–137, *131*, *132*, *133*, *134*, *135*, *136*; benefits of 128–130; design software 139–140; estimating 137–138; geographical information system 143; global navigation satellite system 141–143; global positioning system 141; knowledge application exercises 144; light detection and ranging 141–143; mobile mapping 141–143, *142*; smartphone applications 138–139; summary 143–144
strategic plan 30, 36
supervision 181

Taiichi Ohno 201
takeoffs 59, **69**; *see also* estimating considerations
theories of motivation 180–181
time and materials estimate 70–72, **71**, **72**
Total Quality Management 182–183

Index

Toyota Production System 201–202, 220; 4P Model 203–206, *204*, *206*; *see also* lean manufacturing
TPS *see* Toyota Production System
TQM *see* Total Quality Management
trade associations 224–226, 256, 258
training modules **231**
training programs 188–190, 229–232; K.A.S.H e.© system 230–232; sample training modules **231**; *see also* productivity basics; professional development

value-added time 201, 217, 220; *see also* lean manufacturing

value engineering 247
vision statements 239–240; *see also* leadership

watchdogs 30, **31**
WIP *see* work in progress schedule
working capital 25; profit-based method of budget process 29; *see also* forecasting, operating budget
working conditions 181
work in progress schedule 93

zero-based budgeting 30, 33; *see also* profit-based method of budget development